CALCULUS
FOR THE PRACTICAL WORKER

MATHEMATICS LIBRARY FOR PRACTICAL WORKERS

Arithmetic *for the Practical Worker*
Algebra *for the Practical Worker*
Geometry *for the Practical Worker*
Trigonometry *for the Practical Worker*
Calculus *for the Practical Worker*

A GROUP OF BOOKS THAT MAKE EASY
THE HOME STUDY OF THE WORKING
PRINCIPLES OF MATHEMATICS

CALCULUS
FOR THE PRACTICAL WORKER

4th Edition

J. E. Thompson

MATHEMATICS LIBRARY FOR PRACTICAL WORKERS

 VAN NOSTRAND REINHOLD COMPANY
NEW YORK CINCINNATI TORONTO LONDON MELBOURNE

Manufactured in the United States of America

Published by Van Nostrand Reinhold Company Inc.
135 West 50th Street, New York, N.Y. 10020

Van Nostrand Reinhold Limited
1410 Birchmount Road
Scarborough, Ontario MIP 2E7, Canada

Van Nostrand Reinhold Austrailia Pty. Ltd.
17 Queen Street
Mitcham, Victoria 3132, Australia

Van Nostrand Reinhold Company Limited
Molly Millars Lane
Wokingham, Berkshire, England

15 14 13 12 11 10 9 8 7 6 5 4 2

Library of Congress Cataloging in Publication Data

Thompson, James Edgar, 1892-
    Calculus for the practical worker.

    (Mathematics library for practical workers)
    Includes index.
    1. Calculus.   I. Peters, Max, 1906-
II. Title.   III. Series.
QA303.T44   1981   515   81-11357
ISBN 0-442-28274-5   AACR2

PREFACE TO FOURTH EDITION

The popular demand for this book and for the other four in the MATHE-MATICS LIBRARY FOR PRACTICAL WORKERS has been maintained through three editions and for over fifty years. The publisher hopes that this Fourth Edition will prove as valuable to a new generation of practical workers who must work with mathematics on a regular, practical, non-theoretical basis.

PREFACE TO THIRD EDITION

THE calculus is of paramount importance in the applied sciences and engineering; it is no less important to higher mathematics. In each respect, it is necessary that the student acquire sufficient skill and knowledge of the calculus to use it proficiently, intelligently, and imaginatively. To these ends the treatment used in previous editions of this book has proved markedly successful and such treatment has been maintained in this new edition. However, certain subjects, no longer of great interest or importance, have been dropped. Others have been added: notably, a chapter on Differential Equations, a knowledge of which is necessary today in order that the student may proceed to advanced topics and to fully apply the calculus to the various problems he may encounter in his work.

PREFACE TO FIRST
AND SECOND EDITIONS

THIS book on simplified calculus is one of a series designed by the author and publisher for the reader with an interest in the meaning and simpler technique of mathematical science, and for those who wish to obtain a practical mastery of some of the more usual and directly useful branches of the science without the aid of a teacher. Like the other books in the series it is the outgrowth of the author's experience with students such as those mentioned and the demand experienced by the publisher for books which may be "read as well as studied."

One of the outstanding features of the book is the use of the method of rates instead of the method of limits. To the conventional teacher of mathematics, whose students work for a college degree and look toward the modern theory of functions, the author hastens to say that for their purposes the limit method is the only method which can profitably be used. To the readers contemplated in the preparation of this book, however, the notion of a limit and any method of calculation based upon it always seem artificial and not in any way connected with the familiar ideas of numbers, algebraic symbolism or natural phenomena. On the other hand, the method of rates seems a direct application of the principle which such a reader has often heard mentioned as the extension of arithmetic and algebra with which he must become acquainted before he can perform calculations which involve *changing* quantities. The familiarity of examples of changing quantities in everyday life also makes it a simple matter to introduce the terminology of the calculus; teachers and readers will recall the difficulty encountered in this connection in more formal treatments.

The scope and range of the book are evident from the table of contents. The topics usually found in books on the calculus but not appearing here are omitted in conformity with the plan of the book as stated in the first paragraph above. An attempt has been made to approach the several parts of the subject as naturally and directly as possible, to show as clearly as possible the unity and continuity of the subject as a whole, to show what the calculus "is all about" and how it is used, and to present the material in as simple, straightforward and informal a style as it will permit.   It is hoped thus that the book will be of the greatest interest and usefulness to the readers mentioned above.

The first edition of this book was prepared before the other volumes of the series were written and the arrangement of the material in this volume was not the same as in the others.   In this revised edition the arrangement has been changed somewhat so that it is now the same in all the volumes of the series.   Some changes and additions have been made in the text, but the experience of readers has indicated that the text is in the main satisfactory, and beyond corrections and improvements in presentation these changes are few.   The last section of the book (Article 109) is new, and in this edition a fairly complete table of the more useful integral formulas has been added.   The greatest changes are in the exercises and problems.   These have been increased considerably in number, some of the original exercises have been replaced by better ones, and answers have been provided for all. Some of the new problems bring in up-to-date illustrations and applications of the principles, and it is hoped that all will now be found more useful and satisfactory.

Many of the corrections and improvements in the text are the results of suggestions received from readers of the first edition, and it is hoped that readers of the new edition will call attention to errors or inaccuracies which may be found in the revised text.

J. E. THOMPSON

Brooklyn, N. Y.
October, 1945

CONTENTS

ix

INTRODUCTION

In arithmetic we study numbers which retain always a fixed value (constants). The numbers studied in algebra may be constants or they may vary (variables), but in any particular problem the numbers remain constant *while a calculation is being made*, that is, throughout the consideration of that one problem.

There are, however, certain kinds of problems, not considered in algebra or arithmetic, in which the quantities involved, or the numbers expressing these quantities, are continually changing. Thus, if a weight is dropped and allowed to fall freely, its speed steadily increases; or, if it is thrown directly upward, it first moves more and more slowly and finally stops, then begins to fall, slowly at first and then faster and faster. Its speed is a quantity which continually changes. Again, as the crankshaft of a gas or steam engine turns, the direction of motion of the crank pin, the speed of the crosshead, and both the direction and speed of the connecting rod, are continually changing. The alternating electric current in our house lighting circuits does not have the same strength at any two successive instants. If an aeroplane crosses over a straight road at the instant at which an automobile passes the crossing point, the rate at which the distance between the aeroplane and the automobile changes is at first much greater than at a later time and depends on the height of the aeroplane, the direction and speed of travel of both aeroplane and automobile, and the distance of each from the crossing point at any particular instant which we may wish to consider.

Many such examples could be cited; in fact, such problems form the greater part of those arising in natural phenomena and in engineering. In order to perform the calculations of such problems and even more to study the relations of the various factors entering into them, whether a numerical calculation is made or not, other methods than those of arithmetic and algebra have been developed. The branch of mathematics which treats of these methods is called the *calculus*. Since the calculus has particularly to do with changing quantities it is obvious that one of its fundamental notions or considerations must be that of rates of change of variable quantities, or simply *rates*.

But the calculus does more than to develop methods and rules for solving problems involving changing quantities. It investigates, so to speak, the inner nature of such a quantity, its origin, the parts of which it consists, the greatest and least values which it may have under stated conditions, its relations to other numbers, the relations between the rates of related sets of numbers, and sums of very great numbers of very small quantities. In short, the calculus deals not only with the use of numbers as does arithmetic and with the symbols and methods of writing numbers as does algebra, but also and more particularly with the *nature* and the *variations* of numbers.

In the calculus, continual use is made of one's knowledge of algebra and trigonometry in dealing with equations, angle functions, formulas, transformations, etc., so that these subjects should be studied first.* It is for this reason, and not because it is any more difficult, that calculus is studied after algebra and trigonometry. The ideas involved in the study of rates and of the sums of very large numbers of very small values are not in themselves at all difficult to grasp when one is familiar with the methods of thought and forms of expression used in algebra, trigonometry, etc.

While it cannot be expected that any mathematical subject can be presented in the so-called "popular and non-mathematical" style, the aim in this book is to present the subject in an *informal* manner and with the smallest amount of technical machinery consistent with a fair statement of its meaning and its relation to elementary mathematics in general. The illustrations of its applications are taken from the problems that interest but puzzle us in our everyday observations of the physical world about us, and from applied science, or engineering. Perhaps not all the illustrative problems will be of interest to any one reader; he may make his own selection. Some readers may wish to study the calculus itself further, while others may be interested in its technical applications. References are made, at appropriate places, to books which will be useful to either.

The large amount of detail work which is necessary in some parts of the calculus, and the manner in which the study of the subject is

---

* *Note.*—All the knowledge of algebra and trigonometry necessary for the study of this book may be obtained from the author's "Algebra for the Practical Man" and "Trigonometry for the Practical Man," published by D. Van Nostrand Company, Princeton, N. J.

sometimes approached, have often led to the false idea that the subject is extremely difficult and one to be dreaded and avoided. However, when it is properly approached and handled the calculus will be found to be simple and fascinating and its mastery will provide the highest intellectual satisfaction and great practical utility.

A brief historical note may be of interest. The method of calculation called "the calculus" was first discovered or invented by Isaac Newton, later Sir Isaac, an English mathematician and physicist, the man who discovered the law of gravitation and first explained the motions of the heavenly bodies, the earth, and objects upon and near the earth under the action of gravity. He wrote out his calculus, used it at Cambridge University (England) where he was Professor, and showed it to his friends, in the year 1665–1666. Some ideas and methods of calculation similar to the calculus were known and used in ancient times by Archimedes and others of the early Greek mathematicians, and also by an Italian named Cavalieri who lived in 1598–1647 and a French mathematician named Roberval (1602–1675). None of these men, however, developed the method beyond a few vague procedures or published any complete accounts of their methods. Newton, on the other hand, developed the methods which we shall study in this book in fairly complete form, used it regularly in his mathematical work, and wrote a systematic account of it.

The calculus was also discovered or invented independently by a German mathematician and philosopher, Gottfried Wilhelm Leibnitz, who published his first account of his method in 1676. The friends and followers of Newton and Leibnitz carried on a great controversy concerning the priority of discovery, the friends of each man claiming that he made the discovery first and independently and that the other copied from him. Nowadays it is generally agreed that Newton and Leibnitz each made his discovery or invention independently of the other, and both are given full credit. Many of the formulas and theorems which we shall use in this book were first made known by Newton, but the symbols which we shall use for differentials, derivatives and integrals are those first used by Leibnitz.

After Newton and Leibnitz made their work known the methods of the calculus came into general use among mathematicians and many European workers contributed to the development, perfection, and applications of the calculus in pure and applied science. Among the leaders in this work were the Bernoulli family (Daniel, James and

Jacques) in Switzerland, Colin MacLaurin in Scotland and England, Louis Joseph Lagrange in France, and Leonhard Euler in Switzerland and France. The first complete text books on the calculus were published by Euler in 1755 (the "Differential Calculus") and in 1768 (the "Integral Calculus"). These books contained all that was then known of the calculus and they have greatly influenced later developments and books on the subject.

Chapter 1

FUNDAMENTAL IDEAS.
RATES AND DIFFERENTIALS

1. Rates. The most natural illustration of a rate is that involving motion and time. If an object is moving steadily as time passes, its speed is the distance or space passed over in a specified unit of time, as, for example, 40 miles per hour, 1 mile per minute, 32 feet per second, etc. This speed of motion is the time rate of change of distance, and is found simply by dividing the space passed over by the time required to pass over it, both being expressed in suitable units of measurement. If the motion is such as to increase the distance from a chosen reference point, the rate is taken as positive; if the distance on that same side of the reference point decreases, the rate is said to be negative.

These familiar notions are visualized and put in concise mathematical form by considering a picture or graph representing the motion. Thus in Fig. 1 let the motion take place along the straight

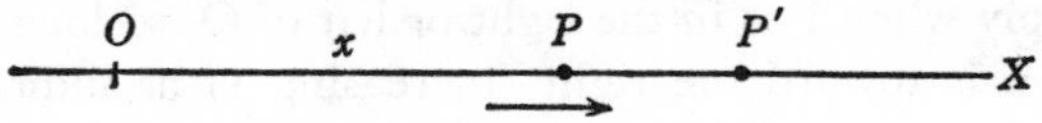

FIG. 1.

line OX in the direction from O toward X. Let O be taken as the reference point, and let P represent the position of the moving object or point. The direction of motion is then indicated by the arrow and the distance of P from O at any particular instant is the length OP which is represented by x.

When the speed is uniform and the whole distance x and the total time t required to reach P are both known, the speed or rate is simply $x \div t$ or $\frac{x}{t}$. If the total distance and time from the starting point are not known, but the rate is still constant, the clock times of passing two points P and P' are noted and the distance between P and P' is measured. The distance PP' divided by the difference in times then gives

1

the rate. Thus even though the distance OP $= x$ or the distance OP′ which may be called x' may not be known, their *difference* which is PP′ $= x' - x$ is known; also the corresponding time difference $t' - t$ is known. Using these symbols, the rate which is *space difference* divided by *time difference*, is expressed mathematically by writing

$$\text{Rate} = \frac{x' - x}{t' - t} = \frac{\text{distance covered}}{\text{time elapsed}}.$$

If the x *difference* is written dx and the t *difference* is written dt then the

$$\text{Rate} = \frac{dx}{dt}. \tag{1}$$

The symbols dx and dt are *not* products d times x or d times t as in algebra, but each represents a single quantity, the x or t difference. They are pronounced as one would pronounce his own initials, thus: dx, "dee-ex"; and dt, "dee-tee." These symbols and the quantities which they represent are called *differentials*. Thus dx is the differential of x and dt the differential of t.

If, in Fig. 1, P moves in the direction indicated by the arrow, the rate is taken as positive and the expression (1) is written

$$\text{Rate} = +\frac{dx}{dt}.$$

This will apply when P is to the right or left of O, so long as the sense of the motion is toward the right (increasing x) as indicated by the arrow. If it is in the opposite sense, the rate is negative (decreasing x) and is written

$$\text{Rate} = -\frac{dx}{dt}.$$

These considerations hold in general and we shall consider always that when the rate of *any* variable is positive the variable is increasing, when negative it is decreasing.

So far the idea involved is familiar and only the terms used are new. Suppose, however, the object or point P is increasing its speed when we attempt to measure and calculate the rate, or suppose it is slowing down, as when accelerating an automobile or applying the brakes to stop it; what is the speed then, and how shall the rate be measured or expressed in symbols? Or suppose P moves on a circle or other curved path so that its direction is changing, and the arrow in Fig. 1 no longer has the significance we have attached to it. How then shall $\dfrac{dx}{dt}$ be measured or expressed?

These questions bring us to the consideration of variable rates and the heart of the methods of calculus, and we shall find that the scheme given above still applies, the key to the question lying in the differentials dx and dt.

The idea of differentials has here been developed at considerable length because of its extreme importance, and should be mastered thoroughly. The next section will emphasize this statement.

2. Varying Rates. With the method already developed in the preceding section, the present subject can be discussed concisely and more briefly. If the speed of a moving point be not uniform, its numerical measure at any particular instant is defined as the number of units of distance which would be described in a unit of time *if the speed were to remain constant from and after that instant.* Thus, if a car is speeding up as the engine is accelerated, we would say that it has a velocity of, say, 32 feet per second *at any particular instant* if it should move for the next second at the same speed it had *at that instant* and cover a distance of 32 feet. The actual space passed over may be greater if accelerating or less if braking, because of the change in the rate which takes place in that second, but the rate *at that instant* would be that just stated.

To obtain the measure of this rate at any specified instant, the same principle is used as was used in article 1. Thus, if in Fig. 1 dt is any chosen interval of time and $PP' = dx$ is the space which would be covered in that interval, *were P to move over the distance PP' with the same speed unchanged which it had at P,* then the rate at

P is $\dfrac{dx}{dt}$. The quantity $\dfrac{dx}{dt}$ is plus or minus according as P moves in the sense of the arrow in Fig. 1 or the opposite.

If the point P is moving on a curved path of any kind so that its direction is continually changing, say, on a circle, as in Fig. 2, then the direction at any instant is that of the tangent to the path at the point P at that instant, as

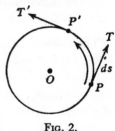

FIG. 2.

PT at P and P'T' at P'. The space differential ds is laid off on the direction *at* P and is taken as the space which P would cover in the time interval dt *if the speed and direction were to remain the same during the interval as at* P. The rate is then, as usual, $\dfrac{ds}{dt}$ and is plus or minus according as P moves along the curve in the sense indicated by the curved arrow or the reverse.

3. Differentials. In the preceding discussions the quantities dx or ds and dt have been called the *differentials* of x, s and t. Now time passes steadily and without ceasing so that dt will always exist. By reference to chosen instants of time the interval dt can be made as great or as small as desired, but it is always formed in the same manner and sense and is always positive, since time never flows backward. The differential of any other variable quantity x may be formed in any way desired if the variation of x is under control and may be great or small, positive or negative, as desired, or if the variable is not under control its differential may be observed or measured and its sense or sign (plus or minus) determined, positive for an increase during the interval dt and negative for a decrease. The rate of x, dx/dt, will then depend on dx and since dt is always positive, dx/dt will be positive or negative according as dx is plus or minus.

From the discussions in articles 1 and 2 it is at once seen that the definition of the differential of a variable quantity is the following:

The differential dx of a variable quantity x at any instant is the change in x which would occur in the next interval of time dt if x were to continue to change uniformly in the interval dt with the same rate which it has at the beginning of dt.

Using this definition of the differential we then define:

The mathematical rate of x at the specified instant is the quotient of dx by dt, that is, the ratio of the differentials.

The differential of *any* variable quantity is indicated by writing the letter d before the symbol representing the quantity. Thus the differential of x^2 is written $d(x^2)$, the differential of $\sqrt{x}$ is written $d(\sqrt{x})$. The differential of x^2 or of $\sqrt{x}$ will of course depend on the differential of x itself. Similarly $d(\sin \theta)$ will depend on $d\theta$, $d(\log_b x)$ will depend on dx and also on the base b. When the differentials $d(x^2)$, $d(\sqrt{x})$, $d(\sin \theta)$ are known or expressions for them have been found then the rates of these quantities will be

$$\frac{d(x^2)}{dt}, \quad \frac{d(\sqrt{x})}{dt}, \quad \frac{d(\sin\theta)}{dt}, \text{ etc.,}$$

and will depend on the rates dx/dt, $d\theta/dt$, etc.

Now, in mathematical problems, such expressions as x^2, $\sqrt{x}$, $\sin\theta$, $\log x$, $x + y$, $x - y$, xy, x/y, etc., are of regular and frequent occurrence. In order to study problems involving changing quantities which contain such expressions as the above, it is necessary to be able to calculate their rates and since the rate is the differential of the expression divided by the time differential dt, it is essential that rules and formulas be developed for finding the differentials of any mathematical expressions. Since the more complicated mathematical expressions are made up of certain combinations of simpler basic forms (sum, difference, product, quotient, power, root, etc.) we proceed to find the differentials of certain of the simple fundamental forms.

The finding or calculation of differentials is called *differentiation* and is one of the most important parts of the subject of calculus, that part of the subject which deals with differentiation and its applications being called the *differential calculus*.

4. Mathematical Expression of a Steady State. In order to obtain rules or formulas for the differentials and rates of such expressions as those given in the preceding article we shall first develop a formula or equation which expresses the distance x of the point P from the reference point O in Fig. 1 at any time t after the instant of starting.

If P is moving in the positive direction at the constant speed k, then we can write that the rate is

$$\frac{dx}{dt} = k. \tag{2}$$

At this speed the point P will, in the length of time t, move over a distance equal to kt, the speed multiplied by the time. If at the beginning of this time, the instant of starting, P were already at a certain fixed distance a from the reference point O, then at the end of the time t the total distance x will be the sum of the original distance a and the distance covered in the time t, that is,

$$x = a + kt. \tag{3}$$

If P starts at the same point and moves in the opposite (negative) direction, then the total distance after the time t is the difference

$$x = a - kt \tag{4}$$

and the rate is

$$\frac{dx}{dt} = -k. \tag{5}$$

The several equations (2) to (5) may be combined by saying that if

$$x = a \pm kt, \quad \frac{dx}{dt} = \pm k. \tag{6}$$

Considering expressions (2) and (3), since x equals $a + kt$ then of course the rate of x equals the rate of $a + kt$, that is, $dx/dt = d(a + kt)/dt$. But by (2) $dx/dt = k$, therefore

$$\frac{d(a + kt)}{dt} = k. \tag{7}$$

In the same manner from (4) and (5) we get

$$\frac{d(a - kt)}{dt} = -k.$$

These results will be used in finding the differentials of other simple expressions. It is to be remembered that equation (3) is the expression for the value of any variable x (in this case a distance) at any time t when its rate is constant, and that (7) gives the value of this rate in terms of the right side of (3), which is equal to the variable x.

5. Differential of a Sum or Difference of Variables. We can arrive at an expression for the differential of a sum or difference of two or more variables in an intuitive way by noting that since the sum is made up of the parts which are the several variables, then, if each of the parts changes by a certain amount which is expressed as its differential, the change in the sum, which is its differential, will of course be the sum of the changes in the separate parts, that is, the sum of the several differentials of the parts. In order to get an exact and logical expression for this differential, however, it is better to base it on the precise results established in the preceding article, which are natural and easily understood, as well as being mathematically correct.

Thus let k denote the rate of any variable quantity x (distance or any other quantity), and k' the rate of another variable y. Then, as in the example of the last article, we can write, as in equation (3),

$$x = a + kt \tag{3}$$

and also
$$y = b + k't$$

the numbers a and b being the constant initial values of x and y. Adding these two equations member by member we get

$$x + y = a + b + kt + k't$$

or

$$(x + y) = (a + b) + (k + k')t.$$

Now, this equation is of the same form as equation (3), $(x + y)$ replacing x and $(a + b)$, $(k + k')$ replacing a, k, respectively. As in (2) and (7), therefore,

$$\frac{d(x + y)}{dt} = (k + k').$$

But k is the rate of x, dx/dt, and k' is the rate of y, dy/dt. Therefore

$$\frac{d(x + y)}{dt} = \frac{dx}{dt} + \frac{dy}{dt}.$$

Multiplying both sides of this equation by dt in order to have differentials instead of rates, there results

$$d(x + y) = dx + dy. \tag{8}$$

If instead of adding the two equations above we had subtracted the second from the first, we would have obtained instead of (8) the result

$$d(x - y) = dx - dy.$$

This result and (8) may be combined into one by writing

$$d(x \pm y) = dx \pm dy. \tag{9}$$

In the same way three or more equations such as (3) above might be written for three or more variables x, y, z, etc., and we would obtain instead of (9) the result

$$d(x \pm y \pm z \pm \cdots) = dx \pm dy \pm dz \pm \cdots, \tag{A}$$

the dots meaning "and so on" for as many variables as there may be.

We shall find that formula (A) in which x, y, z, etc., may be any single variables or other algebraic terms is of fundamental importance and very frequent use in the differential calculus.

6. Differential of a Constant and of a Negative Variable. Since a constant is a quantity which does not change, it has no rate or differential, or otherwise expressed, its rate or differential is zero.   That is, if c is a constant

$$dc = 0. \tag{B}$$

Then, in an expression like $x + c$, since c does not change, any change in the value of the entire expression must be due simply to the change in the variable x, that is, the differential of $x + c$ is equal simply to that of x and we write

$$d(x + c) = dx. \tag{C}$$

This might also have been derived from (8) or (9).   Thus,

$$d(x \pm c) = dx \pm dc$$

but by (B) $dc = 0$ and, therefore,

$$d(x \pm c) = dx,$$

which is the same as (C), either the plus or minus sign applying in (C).

Consider the expression

$$y = -x; \quad \text{then} \quad y + x = 0$$

and

$$d(y + x) = d(0),$$

but zero does not change and therefore $d(0) = 0$.   Therefore,

$$d(y + x) = dy + dx = 0, \quad \text{or} \quad dy = -dx.$$

But $y = -x$; therefore,

$$d(-x) = -dx. \tag{D}$$

7. Differential of the Product of a Constant and a Variable. Let us refer now to formula (A) and suppose all the terms to be the same; then

$$d(x + x + x + \cdots) = dx + dx + dx + \cdots$$

If there are m such terms, with m constant, then the sum of the terms is mx and the sum of the differentials is $m \cdot dx$.   Therefore,

$$d(mx) = mdx. \tag{E}$$

Since we might have used either the plus or minus sign in (A) we may write (E) with either $+m$ or $-m$.   In general, (E) holds good for any constant m, positive or negative, whole, fractional or mixed, and regardless of the form of the variable which is here represented by x.

Chapter 2

FUNCTIONS AND DERIVATIVES

8. Meaning of a Function. In the solution of problems in algebra and trigonometry one of the important steps is the expression of one quantity in terms of another. The unknown quantity is found as soon as an equation or formula can be written which contains the unknown quantity on one side of the equation and only known quantities on the other. Even though the equation does not give the unknown quantity explicitly, if any relation can be found connecting the known and unknown quantities it can frequently be solved or transformed in such a way that the unknown can be found if sufficient data are given.

Even though the data may not be given so as to calculate the numerical value of the unknown, if the connecting relation can be found the problem is said to be solved. Thus, consider a right triangle having legs x, y and hypotenuse c and suppose the hypotenuse to retain the same value (c constant) while the legs are allowed to take on different consistent values (x and y variable). Then, to every different value of one of the legs there corresponds a definite value of the other leg. Thus if x is given a particular length consistent with the value of c, y can be determined. This is done as follows: The relation between the three quantities x, y, c is first formulated. For the right triangle, this is,

$$x^2 + y^2 = c^2. \tag{10}$$

Considering this as an algebraic equation, in order to determine y when a value is assigned to x the equation is to be solved for y in terms of x and the constant c. This gives

$$y = \sqrt{c^2 - x^2}. \tag{11}$$

In this expression, whenever x is given, y is determined and to every value of x there corresponds a value of y whether it be numerically calculated or not. The variable y is said to be a *function* of the variable x. The latter is called the *independent variable* and y is called the *dependent variable*. We can then in general define a function by saying that,

9

When two variables are so related that, when the value of one is given, then one or more values of the other are determined, the second variable is said to be a function of the first.

If each value of the independent variable yields exactly one value of the dependent variable, then the function is said to be single-valued. If each value of the independent variable yields more than one value of the dependent variable, then the function is said to be multiple-valued.

For example, $y = x + 2$ is a single-valued function of x, and $y = \pm\sqrt{x^2 + 3}$ is a multiple-valued function of x. In the latter case, we say that y is a double-valued function of x.

Examples of functions occur on every hand in algebra, trigonometry, mechanics, electricity, etc. Thus, in equation (2), x is a function of t; if $y = x^2$ or $y = \sqrt{x}$, y is a function of x. If in the right triangle discussed above, θ be the angle opposite the side y, then, from trigonometry, $y = c \sin \theta$ and with c constant y is a function of θ. Also the trigonometric or angle *functions* sine, cosine, tangent, etc., are functions of their angle; thus $\sin \theta$, $\cos \theta$, $\tan \theta$ are determined as soon as the value of θ is given. In the mechanics of falling bodies, if a body falls freely from a position of rest then at any time t seconds after it begins to fall it has covered a space $s = 16t^2$ feet and s is a function of t; also when it has fallen through a space s feet it has attained a speed of $v = 8\sqrt{s}$ feet per second, and v is a function of s. If a variable resistance R ohms is inserted in series with a constant electromotive force E volts the electric current I in amperes will vary as R is varied and according to Ohm's Law of the electric circuit is given by the formula $I = E/R$; the current is a function of the resistance.

In general, the study and formulation of relations between quantities which may have any consistent values is a matter of *functional relations* and when one quantity is expressed by an equation or formula as a function of the other or others the problem is solved. The numerical value of the dependent variable can then by means of the functional expression be calculated as soon as numerical values are known for the independent variable or variables and constants.

In order to state that one quantity y is a function of another quantity x we write $y = f(x)$, $y = f'(x)$, $y = F(x)$, etc., each symbol expressing a different form of function. Thus, in equation (11) above we can say that $y = f(x)$, and similarly in some of the other relations given, $s = F(t)$, $I = \Phi(R)$, etc.

If in (11) both x and c are variables, then values of both x and c must be given in order that y may be determined, and y is a function of both x and c. This is expressed by writing $y = f(x, c)$. If in Ohm's Law both E and R are variable, then both must be specified before I can be calculated and I is a function of both, $I = \Phi(E, R)$.

9. Classification of Functions. Functions are named or classified according to their form, origin, method of formation, etc.

Thus the sine, cosine, tangent, etc., are called the *trigonometric* or *angular (angle) functions*. Functions such as x^2, $\sqrt{x}$, $x^2 + y$, $3\sqrt{x} - 2/y$, formed by using only the fundamental algebraic operations (addition, subtraction, multiplication, division, involution, evolution) are called *algebraic functions*. A function such as b^x, where b is a constant and x variable, is called an *exponential function* of x and $\log_b x$ is a *logarithmic function* of x. In other branches of mathematics other functions are met with.

Any function that is not an algebraic function is called a *transcendental function*. In general, trigonometric, exponential, and logarithmic functions are transcendental functions. Transcendental functions are of great importance in both pure and applied mathematics.

Another classification of functions is based on a comparison of equations (10) and (11). In (11) y is given explicitly as a function of x and is said to be an *explicit function* of x. In (10) if x is taken as independent variable then y can be found but as the equation stands the value of y in terms of x is not given explicitly but is simply implied. In this case y is said to be an *implicit function* of x. Explicit or implicit functions may be algebraic or transcendental.

In (11) where $y = \sqrt{c^2 - x^2}$ we can also find $x = \sqrt{c^2 - y^2}$ and x is an *inverse function* of y; similarly if $x = \sin \theta$ then $\theta = \sin^{-1} x$ (read "anti-sine" or "angle whose sine is") is the inverse function. In general if y is a function of x then x is the inverse function of y, and so for any two variables.

10. Differential of a Function of an Independent Variable. If y is a function of x, written

$$y = f(x), \tag{12}$$

then, since a given value of x will determine the corresponding value of y, the rate of y, dy/dt, will depend on both x and the rate dx/dt at any particular instant. Similarly, for the same value of dt, dy will depend on both x and dx.

To differentiate a function is to express its differential in terms of both the independent variable and the differential of the independent variable.   Thus, in the case of the function (12) dy will be a function of both x and dx.

If two expressions or quantities are always equal, their rates taken at the same time must evidently be equal and so also their differentials. An equation can, therefore, be differentiated by finding the differentials of its two members and equating them.   Thus from the equation

$$(x + c)^2 = x^2 + 2cx + c^2$$
$$d[(x + c)^2] = d(x^2) + d(2cx) + d(c^2),$$

by differentiating both sides and using formula (A) on the right side. Since c, c^2 and 2 are constants, then by formulas (B) and (E) $d(2cx) = 2c\,dx$ and $d(c^2) = 0$.   Therefore,

$$d[(x + c)^2] = d(x^2) + 2c\,dx. \qquad (13)$$

Thus, if the function $(x + c)^2$ is expressed as

$$y = (x + c)^2$$

then,

$$dy = d(x^2) + 2c\,dx, \qquad (13a)$$

and, dividing by dt, the relation between the rates is

$$\frac{dy}{dt} = \frac{d(x^2)}{dt} + 2c\frac{dx}{dt}. \qquad (13b)$$

From equation (13a) we can express dy in terms of x and dx when we can express $d(x^2)$ in terms of x and dx.   This we shall do presently.

11. The Derivative of a Function.   If y is any function of x, as in equation (12),

$$y = f(x), \qquad (12)$$

then, as seen above, the rate or differential of y will depend on the rate or differential of x and also on x itself.   There is, however, another important function of x which can be derived from y which does not depend on dx or dx/dt but only on x.   This is true for any ordinary function whatever and will be proven for the general form (12) once for all.   The demonstration is somewhat formal, but in view of the definiteness and exactness of the result it is better to give it in mathematical form rather than by means of a descriptive and intuitive form.

In order to determine the value of y in the functional equation (12), let the independent variable x have a particular value a at a particular instant and let dx be purely arbitrary, that is, chosen at will. Then, even though dx is arbitrary, so also is dt, and, therefore, the rate dx/dt can be given any chosen definite, fixed value at the instant when $x = a$. Let this fixed value of the rate be

$$\frac{dx}{dt} = k'. \tag{14}$$

The corresponding rate of y will evidently depend on the particular form of the function $f(x)$, as, for example, if the function is $(x + c)^2$ the rate dy/dt is given by equation (13b). Therefore, when dx/dt is definitely fixed, so also is dy/dt. Let this value be represented by

$$\frac{dy}{dt} = k''. \tag{15}$$

Now, since both rates are fixed and definite, so also will be their ratio. Let this ratio be represented by k. Then, from (14) and (15),

$$\frac{\left(\dfrac{dy}{dt}\right)}{\left(\dfrac{dx}{dt}\right)} = \frac{k''}{k'} = k.$$

Now, $(dy/dt) \div (dx/dt) = dy/dx$ and, therefore,

$$\frac{dy}{dx} = k.$$

Since k is definite and fixed, while dx may have any arbitrary value, then k cannot depend on dx. That is, the quantity dy/dx, which is equal to k, *cannot depend on dx*. It must depend on x alone, that is, dy/dx is a function of x. In general, it is a new function of x different from the original function $f(x)$ from which it was derived. This *derived function* is denoted by $f'(x)$. We write, then

$$\frac{dy}{dx} = f'(x). \tag{16}$$

This new function is called the *derivative* of the original function $f(x)$.

Since (16) can also be written as

$$dy = f'(x) \cdot dx, \tag{17}$$

in which the differential of the dependent variable is equal to the product of the derivative by the differential of the independent variable, the derivative is also sometimes called the *differential coefficient* of y regarded as a function of x.

There are thus several ways of viewing the function which we have called the derivative. If we are thinking of a function $f(x)$ as a mathematical expression in any form, then the derivative is thought of as the derived function. If we refer particularly to the dependent variable y as an explicit function of the independent variable x, then we express the derivative as dy/dx (read "dy by dx") and refer to it as the "derivative of y with respect to x."

The derivative was first found, however, as the *ratio of the rates* of dependent and independent variables, from equations (14) and (15), and this is its proper definition. Using this definition, that is, $(dy/dt) \div (dx/dt) = f'(x)$, we have

$$\frac{dy}{dt} = f'(x) \cdot \frac{dx}{dt} \tag{18}$$

and for use in practical problems involving varying quantities this is the most useful way of viewing it. Based on this definition, equation (18) tells us that when we once have an equation expressing one variable as a function of another, the derivative is the function or quantity by which the rate of the independent variable must be multiplied in order to obtain the rate of the dependent variable.

A geometrical interpretation of this important function as applied to graphs will be given later.

In order to find this important function in any particular case equation (16) tells us that we must find the *differential* of the dependent variable and divide it by the differential of the independent variable. In the next chapter we take up the important matter of finding the differentials and derivatives of some fundamental algebraic functions.

Chapter 3

DIFFERENTIALS
OF ALGEBRAIC FUNCTIONS

12. Introduction. In the preceding chapter we saw that in order to find the derivative of a function we must first find its differential, and in Chapter 1 we saw that in order to find the rate of a varying quantity, we must also first find its differential. We then found the differentials of a few simple but important forms of expressions. These will be useful in deriving formulas for other differentials and are listed here for reference.

$$d(x \pm y \pm z \pm \cdots) = dx \pm dy \pm dz \pm \cdots \qquad \text{(A)}$$

$$dc = 0 \qquad \text{(B)}$$

$$d(x + c) = dx \qquad \text{(C)}$$

$$d(-x) = -dx \qquad \text{(D)}$$

$$d(mx) = m \, dx \qquad \text{(E)}$$

In Chapter 2 we found that when we have given a certain function of an independent variable, the derivative of the function can be obtained by expressing the differential of the function in terms of the independent variable and its differential, and then dividing by the differential of the independent variable. We now proceed to find the differentials of the fundamental algebraic functions, and it is convenient to begin with the square of a variable.

13. Differential of the Square of a Variable. In order to find this differential let us consider the expression mx of formula (E), and let

$$z = mx, \quad \text{then} \quad z^2 = m^2 x^2$$

by squaring; z being the dependent variable and m being a constant. Differentiating these two equations by formula (E),

$$dz = m \, dx, \quad d(z^2) = m^2 \cdot d(x^2),$$

and dividing the second of these results by the first, member by member,

15

$$\frac{d(z^2)}{dz} = m \cdot \frac{d(x^2)}{dx}.$$

Dividing this result by the original equation $z = mx$ to eliminate the constant m there results

$$\frac{1}{z} \cdot \frac{d(z^2)}{dz} = \frac{1}{x} \cdot \frac{d(x^2)}{dx}. \tag{19}$$

Now, $d(x^2)/dx$ is the derivative of x^2, and similarly for z^2. Furthermore, the connecting constant m has been eliminated and has no bearing on the equation (19). This equation therefore tells us that the derivative of the square of a variable divided by the variable itself (multiplied by the reciprocal) is the same for *any* two variables x and z. It is, therefore, the same for all variables and has a fixed, constant value, say a. Then,

$$\frac{1}{x} \cdot \frac{d(x^2)}{dx} = a.$$

$$\therefore \quad d(x^2) = ax \cdot dx. \tag{20}$$

In order to know the value of $d(x^2)$, therefore, we must determine the constant a. This is done as follows:

Since equation (20) is true for the square of any variable, it is true for $(x + c)^2$ where c is a constant. Therefore,

$$d[(x + c)^2] = a(x + c) \cdot d(x + c).$$

But, by formula (C), $d(x + c) = dx$, hence,

$$d[(x + c)^2] = a(x + c) \cdot dx$$

$$= ax \cdot dx + ac \cdot dx. \tag{21}$$

Also, according to equation (13),

$$d[(x + c)^2] = d(x^2) + 2c \cdot dx,$$

$$= ax \cdot dx + 2c \cdot dx \tag{22}$$

by (20). By (21) and (22), therefore,

$$ax\,dx + ac\,dx = ax\,dx + 2c\,dx,$$

or,

$$ac\,dx = 2c\,dx.$$

$$\therefore \quad a = 2$$

and this value of a in (20) gives, finally,

$$d(x^2) = 2x\,dx. \tag{F}$$

This is the differential of x^2; dividing by dx the derivative of x^2 with respect to x is

$$\frac{d(x^2)}{dx} = 2x. \tag{23}$$

These important results can be stated in words by saying that, "the differential of the square of any variable equals twice the variable times its differential," and, "the derivative of the square of any variable with respect to the variable equals twice the variable."

Referring to article 11, formula (F) corresponds to equation (17) and (23) to equation (16) when $f(x) = x^2$, and therefore $f'(x) = 2x$.

14. Differential of the Square Root of a Variable. Let x be the variable and let

$$y = \sqrt{x}, \quad \text{then} \quad y^2 = x.$$

Differentiating the second equation by formula (F),

$$2y \, dy = dx, \quad \text{or} \quad dy = \frac{dx}{2y}.$$

But $y = \sqrt{x}$, therefore,

$$d(\sqrt{x}) = \frac{1}{2\sqrt{x}} \, dx \tag{G}$$

and the derivative is

$$\frac{d(\sqrt{x})}{dx} = \frac{1}{2\sqrt{x}}. \tag{24}$$

Formula (G) can be put in a somewhat different form which is sometimes useful. Thus, $\sqrt{x} = x^{1/2}$ and

$$\frac{1}{2\sqrt{x}} = \frac{1}{2} \frac{1}{\sqrt{x}} = \frac{1}{2} \frac{1}{x^{1/2}} = \frac{1}{2} x^{-1/2}.$$

Therefore, (G) becomes

$$d(x^{1/2}) = \tfrac{1}{2} x^{-1/2} \, dx. \tag{25}$$

15. Differential of the Product of Two Variables. Let x and y be the two variables. We then wish to find $d(xy)$. Since we already have a formula for the differential of a square we first express the product xy in terms of squares. We do this by writing

$$(x + y)^2 = x^2 + 2xy + y^2.$$

Transposing and dividing by 2, this gives,

$$xy = \tfrac{1}{2}(x + y)^2 - \tfrac{1}{2}x^2 - \tfrac{1}{2}y^2.$$

Differentiating this equation and using formula (A) on the right,

$$d(xy) = d[\tfrac{1}{2}(x + y)^2] - d(\tfrac{1}{2}x^2) - d(\tfrac{1}{2}y^2).$$

Applying formula (F) to each of the squares on the right and handling the constant coefficients by formula (E) we get

$$d(xy) = (x + y) \cdot d(x + y) - x\,dx - y\,dy$$
$$= (x + y)(dx + dy) - x\,dx - y\,dy$$
$$= x\,dx + x\,dy + y\,dx + y\,dy - x\,dx - y\,dy.$$
$$\therefore \quad d(xy) = x\,dy + y\,dx. \tag{H}$$

A simple application of this formula gives the differential of the reciprocal of a variable. Let x be the variable and let

$$y = \frac{1}{x}, \quad \text{then} \quad xy = 1.$$

Differentiating the second equation by formula (H), and remembering that by formula (B) $d(1) = 0$, we get,

$$x\,dy + y\,dx = 0, \quad \text{hence,} \quad dy = -\frac{1}{x} \cdot y\,dx.$$

But, $y = \dfrac{1}{x}$, therefore,

$$d\left(\frac{1}{x}\right) = -\frac{1}{x^2}\,dx \tag{J}$$

is the differential, and the derivative with respect to x is

$$\frac{d\left(\dfrac{1}{x}\right)}{dx} = -\frac{1}{x^2}. \tag{26}$$

Formula (J) can be put into a different form which is often useful. Thus, $1/x = x^{-1}$ and $1/x^2 = x^{-2}$; hence, (J) becomes

$$d(x^{-1}) = -x^{-2}\,dx. \tag{27}$$

16. Differential of the Quotient of Two Variables. Let x, y be the variables; we wish to find $d(x/y)$. Now, we can write x/y as $(1/y) \cdot x$ which is in the form of a product. Applying formulas (H) and (J) to this product, we get

$$d\left(\frac{x}{y}\right) = d\left(\frac{1}{y} \cdot x\right) = \frac{1}{y} \cdot dx + x \cdot d\left(\frac{1}{y}\right)$$
$$= \frac{dx}{y} - x\,\frac{dy}{y^2},$$

or, combining these two last terms with a common denominator,

$$d\left(\frac{x}{y}\right) = \frac{y\,dx - x\,dy}{y^2}. \tag{K}$$

17. Differential of a Power of a Variable. Letting x represent the variable and n represent any constant exponent, we have to find $d(x^n)$. Since a power is the product of repeated multiplication of the same factors, for example, $x^2 = xx$, $x^3 = xxx$, etc., let us consider formula (H):

$$d(xy) = x\,dy + y\,dx.$$

Then

$$\begin{aligned}
d(xyz) &= d[(xy)\cdot z] \\
&= xy\cdot dz + z\cdot d(xy) \\
&= xy\cdot dz + z(y\,dx + x\,dy) \\
&= xy\cdot dz + yz\cdot dx + zx\cdot dy.
\end{aligned}$$

Similarly,

$$d(xyzt) = (xyz)\,dt + (xyt)\,dz + (xzt)\,dy + (yzt)\,dx.$$

Extended to the product of any number of factors, this formula says, "To find the differential of the product of any number of factors multiply the differential of each factor by the product of all the other factors and add the results."

Thus,

$$\begin{aligned}
d(x^3) = d(xxx) &= xx\,dx + xx\,dx + xx\,dx \\
&= 3x^2\,dx = 3x^{3-1}\,dx.
\end{aligned}$$

In the same way

$$d(x^4) = d(xxxx) = 4x^3\,dx = 4x^{4-1}\,dx,$$

$$d(x^5) = 5x^4\,dx = 5x^{5-1}\,dx,$$

and, in general, by extending the same method to any power,

$$d(x^n) = nx^{n-1}\,dx. \tag{L}$$

If $y = x^n$ the derivative is

$$\frac{dy}{dx} = \frac{d(x^n)}{dx} = nx^{n-1}. \tag{28}$$

Referring now to formulas (F), (25), (27) it is seen that they are simply special cases of the general formula (L) with the exponent $n = 2$, $\frac{1}{2}$, -1, respectively. Formula (L) holds good for any value of the exponent, positive, negative, whole number, fractional or mixed.

18. Formulas. The formulas derived in this chapter are collected here for reference in connection with the illustrative examples worked out in the next article.

$$d(x^2) = 2x\,dx \tag{F}$$

$$d(\sqrt{x}) = \frac{1}{2\sqrt{x}}\,dx \tag{G}$$

$$d(xy) = x\,dy + y\,dx \tag{H}$$

$$d\left(\frac{1}{x}\right) = -\frac{1}{x^2}\,dx \tag{J}$$

$$d\left(\frac{x}{y}\right) = \frac{y\,dx - x\,dy}{y^2} \tag{K}$$

$$d(x^n) = nx^{n-1}\,dx \tag{L}$$

19. Illustrative Examples.

1. Find the differential of $x^2 - 2x + 3$.

This is the algebraic sum of several terms, therefore by formula (A) we get for the differential of the entire expression

$$d(x^2) - d(2x) + d(3).$$

By formula (F),

$$d(x^2) = 2x\,dx.$$

By formula (E),

$$d(2x) = 2\,dx.$$

By formula (B),

$$d(3) = 0.$$

Therefore

$$d(x^2 - 2x + 3) = 2x\,dx - 2\,dx$$
$$= 2(x - 1)\,dx.$$

2. Find $d(2x^3 + 3\sqrt{x} - \frac{3}{2}x^2)$.

By formula (A) this is equal to $d(2x^3) + d(3\sqrt{x}) - d(\frac{3}{2}x^2)$.
By (E) and (L),

$$d(2x^3) = 2d(x^3) = 2(3x^2\,dx) = 6x^2\,dx.$$

By (E) and (G),

$$d(3\sqrt{x}) = 3d(\sqrt{x}) = 3\left(\frac{1}{2\sqrt{x}}\,dx\right) = \frac{3}{2}\frac{dx}{\sqrt{x}}.$$

By (E) and (L)

$$d(\tfrac{3}{2}x^2) = \tfrac{3}{2}d(x^2) = \tfrac{3}{2}(2x\,dx) = 3x\,dx.$$

Therefore the required differential is

$$6x^2\,dx + \frac{3}{2}\frac{dx}{\sqrt{x}} - 3x\,dx = 3\left(2x^2 + \frac{1}{2\sqrt{x}} - x\right)dx.$$

3. Differentiate $3xy^2$.

This is the product of x by y^2 with the constant coefficient 3; therefore by formulas (E) and (H) we have

$$d(3xy^2) = 3d(x\cdot y^2)$$

and

$$d(x\cdot y^2) = x\cdot d(y^2) + y^2\cdot d(x) = x\cdot 2y\,dy + y^2\cdot dx = 2xy\,dy + y^2\,dx.$$

Therefore

$$d(3xy^2) = 3(2xy\,dy + y^2\,dx) = 3y(2x\,dy + y\,dx).$$

4. Differentiate $\sqrt{x^2 - 4}$.

By (G),

$$d(\sqrt{x^2 - 4}) = \frac{1}{2\sqrt{x^2 - 4}} \cdot d(x^2 - 4),$$

and by (C) and (F),

$$d(x^2 - 4) = d(x^2) = 2x\,dx$$

$$\therefore\quad d(\sqrt{x^2 - 4}) = \frac{1}{2\sqrt{x^2 - 4}} \cdot 2x\,dx = \frac{x\,dx}{\sqrt{x^2 - 4}}.$$

5. Differentiate $\dfrac{u^2}{y}$.

This is a quotient; therefore, by (K),

$$d\left(\frac{u^2}{y}\right) = \frac{y\cdot d(u^2) - u^2\cdot d(y)}{y^2}$$

$$= \frac{y\cdot 2u\,du - u^2\,dy}{y^2} = \frac{2uy\,du - u^2\,dy}{y^2}$$

$$d\left(\frac{u^2}{y}\right) = \frac{u}{y^2}(2y\,du - u\,dy).$$

6. Differentiate $(x + 2)\sqrt{x^2 + 4x}$.

This is the product of the factors $x + 2$ and $\sqrt{x^2 + 4x}$. Therefore, by formula (H),

$$d[(x + 2)\sqrt{x^2 + 4x}]$$
$$= (x + 2)\cdot d(\sqrt{x^2 + 4x}) + \sqrt{x^2 + 4x}\cdot d(x + 2). \quad \text{(a)}$$

By formula (G),

$$d(\sqrt{x^2 + 4x}) = \frac{1}{2\sqrt{x^2 + 4x}} \cdot d(x^2 + 4x)$$

and

$$d(x^2 + 4x) = d(x^2) + d(4x) = 2x\,dx + 4\,dx = 2(x + 2)\,dx.$$

Therefore,

$$d(\sqrt{x^2 + 4x}) = \frac{2(x + 2)\,dx}{2\sqrt{x^2 + 4x}} = \frac{(x + 2)\,dx}{\sqrt{x^2 + 4x}}. \quad \text{(b)}$$

Also

$$\sqrt{x^2 + 4x}\cdot d(x + 2) = \sqrt{x^2 + 4x}\cdot dx. \quad \text{(c)}$$

Using the results (b) and (c) in expression (a),

$$d[(x + 2)\sqrt{x^2 + 4x}] = (x + 2) \cdot \frac{(x + 2)\,dx}{\sqrt{x^2 + 4x}} + \sqrt{x^2 + 4x}\,dx$$

$$= \left[\frac{(x + 2)^2}{\sqrt{x^2 + 4x}} + \sqrt{x^2 + 4x}\right] dx.$$

This can be simplified, if desired, by writing the two expressions in brackets over the common denominator $\sqrt{x^2 + 4x}$. This gives,

$$\frac{(x + 2)^2 + \sqrt{x^2 + 4x}\cdot\sqrt{x^2 + 4x}}{\sqrt{x^2 + 4x}} = \frac{(x^2 + 4x + 4) + (x^2 + 4x)}{\sqrt{x^2 + 4x}}.$$

Simplifying this last expression we get finally

$$d[(x + 2)\sqrt{x^2 + 4x}] = \frac{2(x^2 + 4x + 2)}{\sqrt{x^2 + 4x}}\,dx.$$

7. Differentiate $(x + y)/(x - y)$.

This is the quotient of $(x + y)$ by $(x - y)$; therefore, by (K)

$$d\left[\frac{(x + y)}{(x - y)}\right] = \frac{(x - y)\cdot d(x + y) - (x + y)\cdot d(x - y)}{(x - y)^2}$$

$$= \frac{(x - y)(dx + dy) - (x + y)(dx - dy)}{(x - y)^2}.$$

Now, by multiplication $[(x - y)(dx + dy)] - [(x + y)(dx - dy)]$ equals

$$[x\,dx + x\,dy - y\,dx - y\,dy] - [x\,dx - x\,dy + y\,dx - y\,dy]$$
$$= x\,dx + x\,dy - y\,dx - y\,dy - x\,dx + x\,dy - y\,dx + y\,dy$$
$$= 2x\,dy - 2y\,dx = 2(x\,dy - y\,dx)$$
$$\therefore \quad d\left(\frac{x+y}{x-y}\right) = \frac{2(x\,dy - y\,dx)}{(x-y)^2}.$$

8. Differentiate $\frac{2}{3}x^{3/2}$

$$d(\tfrac{2}{3}x^{3/2}) = \tfrac{2}{3}\cdot d(x^{3/2}), \text{ and by (L)}$$
$$d(x^{3/2}) = \tfrac{3}{2}x^{3/2-1}\,dx = \tfrac{3}{2}x^{1/2}\,dx = \tfrac{3}{2}\sqrt{x}\,dx$$
$$\therefore \quad d(\tfrac{2}{3}x^{3/2}) = \tfrac{2}{3}\cdot\tfrac{3}{2}\sqrt{x}\,dx = \sqrt{x}\,dx.$$

9. Find the differential of $(x^2 + 2)^3$.

This is a variable $(x^2 + 2)$ raised to the power 3.   Hence by (L)

$$d[(x^2 + 2)^3] = 3(x^2 + 2)^{3-1}\cdot d(x^2 + 2)$$
$$= 3(x^2 + 2)^2\cdot d(x^2)$$
$$= 3(x^2 + 2)^2\cdot 2x\,dx$$
$$= 6x(x^2 + 2)^2\,dx.$$

10. Differentiate $2/(x^2 + 2)$.

This is the same as $2\left(\dfrac{1}{x^2 + 2}\right)\cdot$ Therefore by formula (J)

$$d\left(\frac{2}{x^2 + 2}\right) = 2\cdot d\left[\frac{1}{(x^2 + 2)}\right] = 2\left[-\frac{d(x^2 + 2)}{(x^2 + 2)^2}\right]$$
$$= 2\left[-\frac{2x\,dx}{(x^2 + 2)^2}\right] = -\frac{4x\,dx}{(x^2 + 2)^2}.$$

11. Differentiate $\dfrac{1}{3(y + 3)^3}\cdot$

This is the same as $\dfrac{1}{3}\cdot\dfrac{1}{(y + 3)^3}$ and $\dfrac{1}{(y + 3)^3} = (y + 3)^{-3}$.   Therefore

$$d\left[\frac{1}{3(y + 3)^3}\right] = d[\tfrac{1}{3}(y + 3)^{-3}] = \tfrac{1}{3}d[(y + 3)^{-3}]$$
$$= \tfrac{1}{3}[-3(y + 3)^{-3-1}\cdot d(y + 3)]$$
$$= \tfrac{1}{3}[-3(y + 3)^{-4}\cdot dy] = -(y + 3)^{-4}\,dy$$
$$= -\frac{dy}{(y + 3)^4}.$$

Exercises.

Differentiate each of the following expressions.

1. $x^3 - 2x^2 + \sqrt{x}$.
2. $x^{3/2} - x^{1/2} + 2$.
3. $\frac{1}{5}x^5 + \frac{1}{4}x^4 + \frac{1}{3}x^3 + \frac{1}{2}x^2 + x$.
4. $(7 - 3x)^4$.
5. $(4 - 2x^3)^3$.

6. $\dfrac{4}{(x - 1)^2}$.

7. $\sqrt[3]{1 - x^3}$.

8. $\dfrac{3 - x^3}{2x^2 + 5}$.

9. $\dfrac{2t}{\sqrt{1 + t}} - 4\sqrt{1 + t}$.

10. $(7 + y)\sqrt{3 - 2y} + y^2$.

11. $(\sqrt{2x + 7})^4$.

12. $\dfrac{3 - x}{\sqrt{x^2 - 6x}}$.

Find the derivative of each of the following expressions:

13. $5x^{1/2} + 6$.
14. $4x^{-1} - 7x^{-2}$.
15. $6x^{1/3} - 9x^{2/3}$.

16. $2\sqrt{x} - \dfrac{1}{2\sqrt{x}}$.

17. $\sqrt{3 + 4x}$.
18. $\sqrt[3]{4 - 3x}$.

19. $\dfrac{1}{\sqrt{a^2 - x^2}}$, ($a$ is constant).

20. $\dfrac{a - x}{a + x}$.

Chapter 4

USE OF RATES
AND DIFFERENTIALS
IN SOLVING PROBLEMS

20. Introductory Remarks. In the preceding chapter we have worked out a number of examples illustrating the use of the differential formulas. In those illustrations and the corresponding exercises for the reader, the functions or algebraic expressions were given already formed. In applying the principles of rates and differentials to problems which are simply described without being formulated, however, the function must first be formulated mathematically.

In the present chapter we give the detailed solutions of a number of such problems showing the use and application of rates and differentials in cases in which variable quantities are involved. In these solutions the formulation of one variable as a function of another is shown in full and the differentiation is carried out by the appropriate formula in each case. The formula is not designated by letter as in the preceding chapter, however, and the reader is advised to look up the proper formula and follow out its application step by step in each case. The differentiations used involve only the algebraic formulas so far derived.

In each case the results obtained are interpreted when their significance is not immediately obvious.

21. Illustrative Problems. (1) A kite is 240 feet high and has 250 feet of string out. If the kite moves horizontally 4 miles per hour away from the boy who is flying it, how fast is the string being let out?

Solution. The lettered steps in the solution of this problem will serve as a model in solving other problems in rates.

a. Draw a diagram and label the parts
   Let x = The horizontal distance from the boy to the kite
   Let y = The length of the string let out
b. Set up a relationship between the variables
   In this case, we have a right triangle, and $y^2 = x^2 + 240^2$

c. Differentiate

$$2y\ dy = d(x^2 + 240^2) = 2x\ dx$$

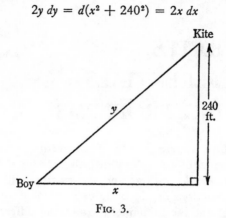

Fig. 3.

d. Divide both sides of the equation by dt

$$2y\ \frac{dy}{dt} = 2x\ \frac{dx}{dt}$$

e. Substitute the known quantities and solve for the unknown
Since $y^2 = x^2 + 240^2$ and $y = 250$, we can find x

$$250^2 = x^2 + 240^2$$
$$62500 = x^2 + 57600$$
$$x^2 = 4900$$
$$x = 70.$$

We also know that $\dfrac{dx}{dt} = 4$

In the equation, $2y\ \dfrac{dy}{dt} = 2x\ \dfrac{dx}{dt}$, we have

$$2(250)\left(\frac{dx}{dt}\right) = 2(70)(4)$$

$$\frac{dy}{dt} = \frac{2(70)(4)}{2(250)} = \frac{28}{25}.$$

Thus, the string is being let out at the rate of $\frac{28}{25}$ miles per hour.

(2) The top of a ladder 20 ft. long is resting against a vertical wall

on a level pavement when the ladder begins to slide downward and outward At the moment when the foot of the ladder is 12 feet from the wall it is sliding away from the wall at the rate of two feet per second. How fast is the top sliding downward at that instant? How far is the foot of the ladder from the wall when it and the top are moving at the same rate?

Solution. In Fig. 4 let OA represent the pavement, OB the wall, and AB the ladder; the arrows represent the direction of motion. Let x

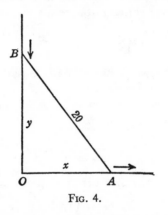

FIG. 4.

represent the distance OA of the foot of the ladder from the wall, and y the distance OB of the top from the pavement. We have then from the statement of the problem AB = 20 feet, $dx/dt = 2$ ft./sec., and we are to find dy/dt and also find when $dy/dt = dx/dt$.

First we must express y in terms of x. This is done from the figure by noting that since OA is horizontal and OB is vertical the triangle AOB is a right triangle. Therefore,

$$\overline{OB}^2 + \overline{OA}^2 = \overline{AB}^2$$

that is,
$$y^2 + x^2 = 20^2 = 400.$$

Therefore,
$$y = \sqrt{400 - x^2}$$

is the desired relation between y and x, that is, the expression of y as a function of x. In order to get the rate dy/dt, we must from this equation find dy and then divide by dt.

Differentiating the equation by the square root formula,

$$dy = d(\sqrt{400 - x^2}) = \frac{d(400 - x^2)}{2\sqrt{400 - x^2}}$$

$$= \frac{-d(x^2)}{2\sqrt{400 - x^2}} = -\frac{2x\,dx}{2\sqrt{400 - x^2}}$$

$$\therefore \quad \frac{dy}{dt} = -\left(\frac{x}{\sqrt{400 - x^2}}\right)\frac{dx}{dt}. \tag{a}$$

Now we had given $dx/dt = 2$ and are to find dy/dt when the distance $x = 12$. Putting these values in the result (a), we get

$$\frac{dy}{dt} = -\frac{12}{\sqrt{400 - (12)^2}} \cdot 2 = -\frac{24}{\sqrt{256}} = -1\tfrac{1}{2} \text{ ft./sec.}$$

The negative sign of dy/dt indicates that y is *decreasing*, that is, the top of the ladder is moving downward.

To find when $dy/dt = dx/dt$, put dy/dt for dx/dt in formula (a). Then the factor dy/dt cancels on each side, and we have

$$1 = -\frac{x}{\sqrt{400 - x^2}}.$$

Hence,

$$400 - x^2 = x^2, \quad 2x^2 = 400, \quad x^2 = 200.$$

$$\therefore \quad x = 14.14 \text{ ft.}$$

That is, at the instant when the foot of the ladder is 14.14 feet from the wall the foot and top are moving at the same rate.

(3) A stone is dropped into a quiet pond and waves move in circles outward from the place where it strikes, at a speed of three inches per second. At the instant when the radius of one of the wave rings is three feet, how fast is its enclosed area increasing?

Solution. Let R be the radius and A the area of one of the circular waves. Then $A = \pi R^2$ and $dA = \pi \cdot d(R^2) = 2\pi R\,dR.$

$$\frac{dA}{dt} = 2\pi R \frac{dR}{dt}.$$

The speed of the wave outward from the center is the rate at which the radius increases, dR/dt. Hence, $dR/dt = 3\,\dfrac{\text{in.}}{\text{sec.}} = \dfrac{1}{4}\dfrac{\text{ft.}}{\text{sec.}}$ and at the instant when the radius is $R = 3$ feet the area is increasing at the rate

$$\frac{dA}{dt} = 2\pi \cdot 3 \cdot \frac{1}{4} = \frac{3\pi}{2} = 4.71 \text{ sq. ft./sec.}$$

This problem can be formulated graphically by means of Fig. 5. As the circular wave moves from the inner dotted circle to the very

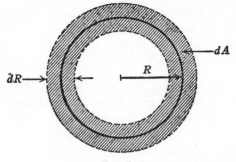

Fig. 5.

near outer one, the differential of the radius is dR and the corresponding differential increase of area is the shaded ring of area dA.

Now the average radius of this ring strip is R and, therefore, its length is $2\pi R$. Its area is the product of length by width.

$$dA = 2\pi R \cdot dR,$$

which is the result already obtained by differentiation. From this result, the rate is found as before.

(4) Water runs into a conical paraffine paper cup five inches high and three inches across the top, at the rate of one cubic inch per second. When it is just half filled how rapidly is the surface of the water rising?

Solution. Let Fig. 6 represent the shape and dimensions of the cup. Then, if v represents the volume of water already in the cup when h is the height of the surface above the point of the cone, the rate at which water is running in is the rate of increase of the volume, dv/dt, and dh/dt is the rate of increase of the

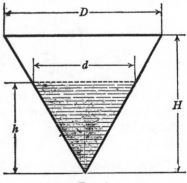

Fig. 6.

height h, that is, the rate at which the surface is rising. The indicated dimensions d and h (this d has nothing to do with differentiation) are then variables and D, H are constants. We have $dv/dt = 1$ cu. in. per sec. and are to find dh/dt at the instant when h is such that the cup is half filled.

The volume of the cone is one third the base area times the altitude. Hence, the total volume is

$$\frac{1}{3}\left(\frac{\pi D^2}{4}\right) \cdot H = \frac{\pi D^2 H}{12} = \frac{\pi (3^2) \cdot 5}{12} = 11.7,$$

with the dimensions given. Therefore, when half filled, the volume of the water is $v = 5.85$ cu. in. We must, therefore, find dh/dt when $v = 5.85$ and $dv/dt = 1$, and to do this we must have a relation between h and v, that is, express h as a function of v.

The volume formula in terms of the altitude h furnishes the relation desired. This formula is, as above,

$$v = \frac{\pi d^2 h}{12}, \quad \text{hence,} \quad hd^2 = \frac{12}{\pi} v. \tag{a}$$

This formula, however, contains not only the desired variables h and v, but also the undesired variable d. This variable must, therefore, be expressed in terms of either h or v. It is simpler to express d in terms of h by means of the proportionality between D, H, which are known, and d, h. In Fig. 6 the two inverted triangles of bases (diameters) D, d and heights H, h are similar and therefore $d:h::D:H$. Therefore,

$$d/h = D/H = \tfrac{3}{5} = .6$$
$$\therefore \quad d = .6h, \quad \text{and,} \quad d^2 = .36h^2.$$

Using this value of d^2 in formula (a) above, it becomes

$$.36h^3 = \frac{12}{\pi} v, \quad \text{hence,} \quad h = \sqrt{\frac{12}{.36\pi}} \cdot \sqrt[3]{v}.$$

By taking the cube root of the numerical fraction and expressing the cube root of v as the $\tfrac{1}{3}$ power, we get finally as the desired functional relation between h and v,

$$h = 2.2v^{1/3}. \tag{b}$$

Differentiating by the power formula,

$$dh = 2.2d(v^{1/3}) = 2.2(\tfrac{1}{3}v^{1/3-1}\,dv) = .74v^{-2/3}\,dv$$

$$\therefore \quad dh = \frac{.74}{\sqrt[3]{v^2}}\,dv, \quad \frac{dh}{dt} = \frac{.74}{\sqrt[3]{v^2}} \cdot \frac{dv}{dt}.$$

Therefore, when $v = 5.85$ (half filled) and $dv/dt = 1$ (rate of inflow),

$$dh/dt = .74/\sqrt[3]{(5.85)^2} = .74/\sqrt[3]{34.2} = .23 \text{ in./sec.}$$

is the rate at which the surface of the water is rising.

This problem illustrates a condition which is often met in calculus: the algebra and geometry or other calculations which are necessary for the formulation before the calculus can be applied are longer than the direct solution of the calculus problem itself. Thus in this case after h was expressed as a function of v in formula (b) the differentiation and calculation of the rate were simple operations. The tedious part of the solution of the problem consisted not in the application of the calculus, but in deriving the functional relation (b) and in calculating v when the cup was half filled.

(5) A ship is sailing due north at the rate of 20 miles per hour. At a certain time another ship crosses its route 40 miles north sailing due east 15 miles per hour. (i) At what rate are the ships approaching or separating after one hour? (ii) After two hours? (iii) After how long are they momentarily neither approaching nor separating? (iv) At that time, how far apart are they?

Solution. We must express the distance between the ships as a function of the time after the second crossed the path of the first. The rate of change of this distance is then their speed of approach or separation. In Fig. 7, let P represent the position of the first ship when the second crosses its path at O, 40 miles due north. After a certain time t hours the ship sailing east will have reached a point A, and the ship sailing north will have reached a point B. The distance between them is then AB. This distance is to be expressed as a function of the time t since A passed O and B left P.

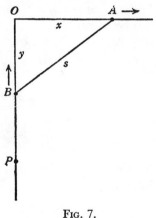

Fig. 7.

If we take O as reference point, and let $OA = x$, $OB = y$, $AB = s$, then $OP = 40$ and

$$s = \sqrt{x^2 + y^2}. \tag{a}$$

Now, the rate of the ship B is $dy/dt = 20$ and of the ship A $dx/dt = 15$

miles per hour. Therefore, after the time t hours has passed, B has covered the distance $\overline{PB} = 20 \cdot t$ and the ship A the distance $\overline{OA} = 15 \cdot t$. Then, $\overline{OB} = \overline{OP} - \overline{PB} = 40 - 20 \cdot t$. Therefore,

$$x = 15t, \quad y = 40 - 20t. \tag{b}$$

Using these values of x, y in equation (a) gives for the distance

$$s = \sqrt{(15t)^2 + (40 - 20t)^2} = \sqrt{625t^2 - 1600t + 1600}$$

$$\therefore \quad s = 5\sqrt{25t^2 - 64t + 64}. \tag{c}$$

This is the desired relation between the distance s between the ships, and the time t after the crossing at O. If at any time the rate ds/dt is positive, the distance is increasing, that is, the ships are separating. If at any time it is negative they are approaching. To find the rate ds/dt we must differentiate equation (c). Using the square root formula and taking account of the constant multiplier 5, we get

$$ds = 5\left[\frac{d(25t^2 - 64t + 64)}{2\sqrt{25t^2 - 64t + 64}} \right].$$

Differentiating the expression in the numerator of this fraction by the sum rule and each of the individual terms by the appropriate formula, this becomes:

$$ds = \frac{5}{2} \frac{d(25t^2) - d(64t) + d(64)}{\sqrt{25t^2 - 64t + 64}}$$

$$= \frac{5}{2} \frac{50t \, dt - 64 \, dt}{\sqrt{25t^2 - 64t + 64}} = \frac{5(25t - 32) \, dt}{\sqrt{25t^2 - 64t + 64}}$$

$$\therefore \quad \frac{ds}{dt} = \frac{5(25t - 32)}{\sqrt{25t^2 - 64t + 64}}. \tag{d}$$

In order to calculate the required results (i) to (iv) we proceed as follows:

(i) After 1 hour $t = 1$ and $ds/dt = 5(25 - 32)/\sqrt{25 - 64 + 64}$.

$\therefore$ $ds/dt = -7$ mi./hr. and the ships are approaching.

(ii) After 2 hours $t = 2$ and $ds/dt = 5(50 - 32)/\sqrt{100 - 128 + 64}$.

$\therefore$ $ds/dt = +15$ mi./hr. and they are separating.

(iii) At the instant when they just cease to approach and begin to separate they are at their nearest positions and momentarily are neither approaching nor separating and $ds/dt = 0$.

$$\therefore \quad \frac{5(25t - 32)}{\sqrt{25t^2 - 64t + 64}} = 0$$

and for this this fraction to equal zero the numerator must be zero.

$$\therefore \quad 5(25t - 32) = 0, \quad \text{or} \quad 25t - 32 = 0$$

$$\therefore \quad t = 1\tfrac{7}{25} \text{ hours} = 1 \text{ hr. } 16 \text{ min. } 48 \text{ sec.}$$

(iv) After a time $t = 1\tfrac{7}{25}$ hours, equations (b) give

$$x = 96/5 \text{ mi.,} \quad y = 72/5 \text{ mi.}$$

The distance between the ships is then according to equation (a)

$$s = \sqrt{(\tfrac{96}{5})^2 + (\tfrac{72}{5})^2} = 24 \text{ mi.}$$

Or, directly by equation (c),

$$s = \sqrt{25(1\tfrac{7}{25})^2 - 64(1\tfrac{7}{25}) + 64} = 24 \text{ mi.}$$

This completes the solution.   Section (iii) of this problem illustrates a use of the derivative which we shall find to be very important.

(6) A helicopter flying horizontally in a straight line at a rate of 60 miles an hour and an elevation of 1760 feet crosses at right angles a straight level road just as an automobile passes underneath at 30 miles per hour. How far apart are they and at what rate are they separating one minute later?

Solution.   In Fig. 8 let C represent the position of the helicopter at

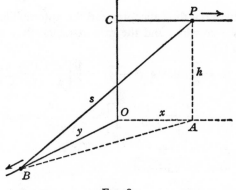

FIG. 8.

the instant when the automobile is vertically below it at the point O. Then CP is the direction of the helicopter and OB that of the automobile. If the arrows indicate the motion, then after a time t minutes they will occupy the position P and B and the straight line distance between them is the inclined diagonal PB = s. We have to find an expression for s at any time t and also its rate ds/dt.

Draw OA parallel to CP; draw the vertical line PA to the point vertically under P; draw AB; and denote the distances by x, y, h as shown. Then OAB is a right triangle with legs OA and OB and hypotenuse AB, and PAB is a right triangle with legs AP and AB and hypotenuse PB = s. Therefore,

$$s^2 = \overline{AB}^2 + h^2, \quad \text{and} \quad \overline{AB}^2 = x^2 + y^2.$$

$$\therefore \quad s^2 = x^2 + y^2 + h^2. \tag{a}$$

Since the helicopter is travelling at the rate $dx/dt = 60$ miles per hour = 1 mi./min. and the automobile at the rate $dy/dt = 30$ miles per hour = $\frac{1}{2}$ mi./min., then, at the end of t minutes they are at the distances from the crossing point of

$$\text{CP} = x = t, \quad \text{OB} = y = \tfrac{1}{2}t, \quad \text{OC} = h = \tfrac{1}{3} \text{ miles}, \tag{b}$$

since 1760 feet is one third of a mile. Using these values of x, y, h in (a) we have

$$s^2 = t^2 + \left(\frac{1}{2}t\right)^2 + \left(\frac{1}{3}\right)^2 = \frac{5t^2}{4} + \frac{1}{9} = \frac{45t^2 + 4}{36}.$$

$$\therefore \quad s = \tfrac{1}{6}\sqrt{45t^2 + 4} \tag{c}$$

is the distance between the helicopter and the automobile at any time t minutes after the crossing, and the rate at which they are separating is ds/dt. From (c)

$$ds = \frac{1}{6} \cdot d(\sqrt{45t^2 + 4}) = \frac{1}{6}\left[\frac{d(45t^2 + 4)}{2\sqrt{45t^2 + 4}}\right]$$

$$= \frac{1}{6}\left[\frac{90t\,dt}{2\sqrt{45t^2 + 4}}\right] = \frac{15t}{2\sqrt{45t^2 + 4}}\,dt.$$

$$\therefore \quad \frac{ds}{dt} = \frac{15t}{2\sqrt{45t^2 + 4}} \tag{d}$$

is the rate at which they are separating at the time t.

Using (c) and (d) we are to calculate the distance and the rate s, ds/dt at the end of one minute. Thus $t = 1$ and by (c)

$$s = \tfrac{1}{6}\sqrt{45 + 4} = \tfrac{7}{6} \text{ miles};$$

by (d),

$$\frac{ds}{dt} = \frac{15}{2\sqrt{45 + 4}} = \frac{15}{14}\frac{\text{mi.}}{\text{min.}} = 64\frac{2}{7} \text{ mi./hr.}$$

In the same way formulas (c) and (d) will give the distance and relative velocity of the two at any other time in minutes.

22. Problems for Solution. The following problems are in general similar to those solved and explained in the preceding article. In each case the quantity or variable whose rate is to be found is to be expressed as a function of the variable or variables whose rates are known. Differentiation by the appropriate formula and division by dt will then give the formula involving the rates, from which the desired result is to be obtained.

Suggestions are given in some cases and a figure should be drawn whenever the nature of the problem will allow. Care must be used in the algebraic formulations and transformations.

Problems.

1. Air is blown into a spherical rubber balloon at such a rate that the radius is increasing at the rate of one-tenth inch per second. At what rate is the air being blown in when the radius is two inches?

(*Hint:* The required rate is dV/dt, where V is the volume.)

2. A metal plate in the shape of an equilateral triangle is being heated in such a way that each of the sides is increasing at the rate of ten inches per hour. How rapidly is the area increasing at the instant when each side is 69.28 inches?

3. Two points move, one on the OX axis and one on OY, in such a manner that in t minutes their distances from O are

$$x = 2t^2 - 6t, \quad y = 6t - 9$$

feet. (i) At what rate are they approaching or separating after one minute? (ii) After three minutes? (iii) When will they be nearest together?

(*Hint:* Find s and ds/dt. In (iii) put $ds/dt = 0$.)

4. A man whose height is six feet walks directly away from a lamp post at the rate of three miles an hour on a level pavement. If the lamp is ten feet above the pavement, at what rate is the end of his shadow travelling?

(*Suggestion:* Draw a figure and denote the variable distance of the man from the post by x, that of the end of the shadow by y, and express y as a function of x by similar triangles.)

5. At what rate does the shadow in Prob. 4 increase in length?

6. A man is walking along the straight bank of a river 120 feet wide toward a boat at the bank, at a rate of five feet a second. At the moment when he is still fifty feet from the boat how rapidly is he approaching the point on the opposite bank directly across from the boat?

(Draw figure, let x = distance to boat, and formulate distance to point opposite.)

7. A man standing on a wharf is hauling in a rope attached to a boat, at the rate of four feet a second. If his hands are nine feet above the point of attachment how fast is the boat approaching the wharf when it is twelve feet away?

8. One end of a wire wound on a reel is fastened to the top of a pole 35 feet high; two men holding the reel on a rod on their shoulders five feet above the level ground walk away from the pole at the rate of five miles an hour, keeping the wire straight. How far are they from the pole when the wire is unwinding at the rate of one mile an hour?

9. A three-mile wind blowing on a level is carrying a kite directly away from a boy. How high is the kite when it is directly over a point 100 feet away and he is paying out the string at the rate of 88 feet a minute?

10. Two automobiles are moving along straight level roads which cross at an angle of sixty degrees, one approaching the crossing at 25 miles an hour and the other leaving it at 30 miles an hour on the same side. How fast are they approaching or separating from each other at the moment when each is ten miles from the crossing?

11. Assuming the volume of a tree to be proportional to the cube of its diameter ($V = k \cdot D^3$ where k is a constant) and that the diameter increases always at the same rate, how much more rapidly is the tree growing in volume when the diameter is three feet than when it is six inches?

12. In being heated up to the melting point, a brick-shaped ingot of silver expands the thousandth part of each of its three dimensions for each degree temperature increase. At what rate per degree (dV/dT, where T is the temperature) is its volume increasing when the dimensions are $2 \times 3 \times 6$ inches?

13. Sand is being poured from a dumping truck and forms a conical pile with its height equal to one third the base diameter. If the truck is emptying at the rate of 720 cubic feet a minute and the outlet is five feet above the ground, how fast is the pile rising as it reaches the outlet?

14. A block of building stone is to be lifted by a rope 50 ft. long passing over a pulley on a window ledge 25 feet above the level ground. A man takes hold of the loose end of the rope which is held five feet above the ground and walks away from the block at ten feet a second. How rapidly will the block *begin* to rise?

15. The volume of a sphere is increasing at the rate of 16 cu. in. per second. At the instant when the radius is 6 in. how fast is it increasing?

16. A rope 28 feet long is attached to a block on level ground and runs over a pulley 12 feet above the ground. The rope is stretched taut and the free end is drawn directly away from the block and pulley at the rate of 13 ft. per

sec.   How fast will the block be moving when it is 5 feet away from the point directly below the pulley?

17. A tank is in the form of a cone with the point downward, and the height and diameter are each 10 feet.   How fast is the water pouring in at the moment when it is 5 feet deep and the surface is rising at the rate of 4 feet per minute?

18. The hypotenuse AB of a right triangle ABC remains 5 inches long while the other two sides change, the side AC increasing at the rate of 2 in. per min. At what rate is the area of the triangle changing when AC is just 3 inches?

19. A spherical barrage balloon is being inflated so that the volume increases uniformly at the rate of 40 cu. ft. per min.   How fast is the surface area increasing at the moment when the radius is 8 feet?

20. A fighter plane is flying in a straight line on a level course to cross the course of a bomber which is also flying on a level course in a straight line. The fighter is at a level 500 feet above the bomber, and their courses cross at an angle of 60 degrees.   Both planes are headed toward the crossing of their courses and on the same side of it, the bomber flying at 200 miles per hour and the fighter at 300.   At the moment when the fighter is 10 miles and the bomber 7 miles from the crossing point, how rapidly are they approaching one another in a straight line joining the two planes?

Chapter 5

DIFFERENTIALS OF TRIGONOMETRIC FUNCTIONS

23. Angle Measure and Angle Functions. If through the center O of the circle in Fig. 2 (article 2) we lay off the horizontal line OX of Fig. 1 (article 1), join the points O and P, and draw through O the vertical OY, we get Fig. 9. Figure 9 is thus a combination of Figs. 1 and 2 and there are differentials to be measured horizontally, vertically and tangentially.

Motion measured parallel to OX toward the right is to be taken as positive and toward the left as negative, as in Fig. 1; similarly motion parallel to OY and upward is positive, downward is negative. Also as in Fig. 2, if the point P moves in the direction opposite to that of the end of the hand of a clock (counter clockwise), the differential of length along the tangent PT is taken as positive, the opposite sense (clockwise) as negative. As in Fig. 1, horizontal distances are denoted by x, similarly vertical distances are denoted by y. Distances measured along the circumference of the circle are denoted by s and differentials of s, taken along the instantaneous tangent at any point P, by ds, as in Fig. 2.

With the above system of notation x, y are then called the *coordinates* of the point P, and if we imagine the point P to move along the tangent line PT *with the direction and speed it had at* P, covering the differential of distance ds in the time dt as in Fig. 2, the coordinate y will change by the differential amount dy in the positive sense and x will change by dx in the negative sense. If the angle AOP be represented by the symbol θ (pronounced "theta") then as P moves along PP′ the radius OP = a will turn about the center O and θ will increase by the positive differential of angle $d\theta$.

The question we now have to answer is this: When P moves through a space differential ds and the angle θ changes by the corresponding differential $d\theta$, what are the corresponding differentials of the angle functions of θ, that is, what are the values of the differentials $d(\sin \theta)$, $d(\cos \theta)$, etc.?

Since sin θ, cos θ, tan θ, etc., are functions of the independent variable θ, then as we have already seen their differentials will depend on $d\theta$ and also on some other function of θ itself. In the present chapter we shall find the formulas giving these differentials.

In order to find the differentials of the angle functions, or *angular* functions (also called *circular* and *trigonometric* functions, for obvious reasons) we need to understand the method of measurement of θ and $d\theta$. If the angle is expressed in degree measure, then the unit of angle is such that the arc intercepted on the circumference by a central

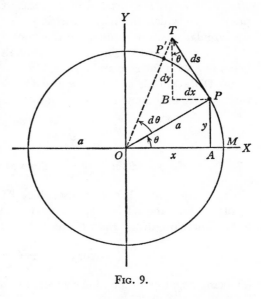

Fig. 9.

angle of one *degree* is equal in length to the 360th part of the circumference. This is the system used in ordinary computation and in the trigonometric tables.

A more convenient system for formulation and analysis is the so-called *circular* measure, in which the unit of angle is such that the arc intercepted on the circumference by a central angle of one *radian* is equal in length to the radius of the circle. Then since the entire circumference equals 2π times the radius, it equals 2π times the arc of one radian and there are 2π radians of angle in the circle. Thus, if the radius of the circle is a, one radian intercepts an arc of length a, and an angle of θ radians intercepts an arc of θ times a. If, therefore, the point P in Fig. 9 moves a distance along the circumference of MP $= s$ while the line OP turns through an angle θ the length of MP is

$$s = a \cdot \theta. \tag{29}$$

By the use of the circular measure we have that $360° = 2\pi$ radians, $180° = \pi$ and $90° = \frac{1}{2}\pi$, with corresponding conversions for other angles. These values will be found more convenient than the values expressed in degrees, and the formula (29) is a much more convenient formula for the length of an arc than the corresponding formula in which the angle is expressed in degrees. Except for numerical computation in which the trigonometric tables have to be used we shall use circular measure throughout this book.*

We now proceed to find the differentials of the angular functions, beginning with the sine and cosine.

24. Differentials of the Sine and Cosine of an Angle. In Fig. 9 we have, by trigonometry, in the right triangle AOP,

$$\sin \theta = \frac{y}{a}, \qquad \cos \theta = \frac{x}{a}$$

$$d(\sin \theta) = \frac{dy}{a}, \quad d(\cos \theta) = \frac{dx}{a}, \tag{30}$$

since a is constant. We have also, since PT is a tangent and perpendicular to OP, and BT is perpendicular to OA, the angle PTB is equal to θ. Therefore in the right triangle PTB, by trigonometry,

$$\frac{(dy)}{(ds)} = \cos \theta, \quad \frac{-(dx)}{(ds)} = \sin \theta$$

(dx being negative in the figure as already pointed out). Hence

$$dy = (\cos \theta) \cdot ds, \quad dx = -(\sin \theta) \cdot ds. \tag{31}$$

If we now go back to equation (29) and differentiate it by formula (E), a being constant, we get

$$ds = a\, d\theta$$

and this value of ds substituted in the two equations (31) gives

$$dy = (\cos \theta) \cdot a\, d\theta, \quad dx = -(\sin \theta) \cdot a\, d\theta,$$

or

$$\frac{dy}{a} = \cos \theta \cdot d\theta, \qquad \frac{dx}{a} = -\sin \theta \cdot d\theta.$$

Substituting these values of dy/a and dx/a in equations (30) we have finally,

$$d(\sin \theta) = \cos \theta \cdot d\theta \tag{M}$$

$$d(\cos \theta) = -\sin \theta \cdot d\theta. \tag{N}$$

Thus the answer to our question in the preceding article is that the differential of the sine of an angle is the differential of the angle multi-

* A detailed explanation of this system of angle measure is given in the author's "Trigonometry for the Practical Man," published by D. Van Nostrand Company, New York, N. Y.

plied by the cosine, and the differential of the cosine is the differential of the angle multiplied by the negative of the sine.

From formulas (M) and (N), the rates of the sine and cosine are,

$$\frac{d(\sin \theta)}{dt} = \cos \theta \frac{d\theta}{dt}, \quad \frac{d(\cos \theta)}{dt} = -\sin \theta \frac{d\theta}{dt} \tag{32}$$

and the derivatives with respect to θ are,

$$\frac{d(\sin \theta)}{d\theta} = \cos \theta, \quad \frac{d(\cos \theta)}{d\theta} = -\sin \theta. \tag{33}$$

25. Differentials of the Tangent and Cotangent of an Angle.
The differential of the tangent is found by applying the fraction formula (K) to the following expression, which is obtained from trigonometry:

$$\tan \theta = \frac{\sin \theta}{\cos \theta}.$$

Differentiating this by the formula (K) we get

$$d(\tan \theta) = d\left(\frac{\sin \theta}{\cos \theta}\right) = \frac{\cos \theta \cdot d(\sin \theta) - \sin \theta \cdot d(\cos \theta)}{\cos^2 \theta}$$

$$= \frac{\cos \theta \cdot (\cos \theta \, d\theta) - \sin \theta \cdot (-\sin \theta \, d\theta)}{\cos^2 \theta}$$

$$= \frac{(\cos^2 \theta + \sin^2 \theta) \, d\theta}{\cos^2 \theta} = \frac{1}{\cos^2 \theta} \, d\theta.$$

since $\cos^2 \theta + \sin^2 \theta = 1$. Also, $1/\cos \theta = \sec \theta$; therefore,

$$d(\tan \theta) = \sec^2 \theta \, d\theta. \tag{P}$$

The differential of cot θ may be found by using the relation from trigonometry $\cot \theta = \cos \theta/\sin \theta$, and proceeding as in the case of the tangent. The following method is, perhaps, shorter. From trigonometry, $\cot \theta = 1/\tan \theta$; therefore, by (J) and (P),

$$d(\cot \theta) = d\left(\frac{1}{\tan \theta}\right) = -\frac{1}{\tan^2 \theta} \cdot d(\tan \theta)$$

$$= -\frac{1}{\tan^2 \theta} \cdot \sec^2 \theta \, d\theta = -\left(\frac{\cos \theta}{\sin \theta}\right)^2 \left(\frac{1}{\cos \theta}\right)^2 d\theta$$

$$= -\left(\frac{1}{\sin^2 \theta}\right) d\theta$$

$$\therefore \quad d(\cot \theta) = -\csc^2 \theta \, d\theta, \tag{Q}$$

since $\dfrac{1}{\sin \theta} = \csc \theta.$

26. Differentials of the Secant and Cosecant of an Angle. Since

$$\sec \theta = \frac{1}{\cos \theta}, \; d(\sec \theta) = -\frac{1}{\cos^2 \theta} \cdot d(\cos \theta)$$

$$= -\frac{1}{\cos^2 \theta} \left(-\sin \theta \; d\theta\right) = \frac{1}{\cos \theta} \cdot \frac{\sin \theta}{\cos \theta} d\theta,$$

or,

$$d(\sec \theta) = \sec \theta \tan \theta \; d\theta. \tag{R}$$

Similarly,

$$d(\csc \theta) = d \left(\frac{1}{\sin \theta}\right) = -\frac{1}{\sin^2 \theta} \cdot d(\sin \theta)$$

$$= -\frac{1}{\sin^2 \theta} \cdot \cos \theta \; d\theta = -\frac{1}{\sin \theta} \cdot \frac{\cos \theta}{\sin \theta} d\theta.$$

$$\therefore \quad d(\csc \theta) = -\csc \theta \cot \theta \; d\theta. \tag{S}$$

This completes the list of differential formulas for the usual trigonometric or angular functions. There are several other angular functions which are of use in certain special branches of applied mathematics, but these are not useful in ordinary work and we will not consider them here. We give next the solution of some examples showing the use and applications of the above formulas.

27. Illustrative Examples Involving the Trigonometric Differentials. In this and the following articles the use of the differential formulas (M), (N), (P), (Q), (R), (S) derived in this chapter will be illustrated by applying them to a few simple examples and problems. Whenever necessary the previous formulas (A) to (L), whose uses have already been illustrated, will be used without referring to them by letter.

1. Find the differential of $3 \sin \theta + a \cos \theta$.

Solution. This expression being the sum of two terms, and 3, a being constants, we have

$$d(3 \sin \theta + a \cos \theta) = 3 \cdot d(\sin \theta) + a \cdot d(\cos \theta)$$

$$= 3(\cos \theta \; d\theta) + a(-\sin \theta \; d\theta)$$

$$= (3 \cos \theta - a \sin \theta) \; d\theta.$$

2. Differentiate $\sin 2\theta$.

Solution. By formula (M), 2 being constant,

$$d(\sin 2\theta) = \cos (2\theta) \cdot d(2\theta) = \cos (2\theta) \cdot 2d\theta = 2 \cos 2\theta \; d\theta.$$

Art. 27 TRIGONOMETRIC FUNCTIONS 43

3. Differentiate $\sin(x^3)$.

Solution. By formula (M), x^3 replacing θ, we get

$$d[\sin(x^3)] = \cos(x^3) \cdot d(x^3), \quad \text{and} \quad d(x^3) = 3x^2\, dx$$

$\therefore \quad d[\sin(x^3)] = 3x^2 \cos x^3\, dx.$

4. Find $d(\cos \sqrt{x})$.

Solution. By formula (N), $\sqrt{x}$ replacing θ, we get

$$d(\cos \sqrt{x}) = -\sin \sqrt{x} \cdot d(\sqrt{x}) = -\sin \sqrt{x} \cdot \frac{dx}{2\sqrt{x}} = -\frac{\sin \sqrt{x}}{2\sqrt{x}}\, dx.$$

5. Find $d[\tan(2x^2 + 3)]$.

Solution. By formula (P), $(2x^2 + 3)$ replacing θ, we get

$$d[\tan(2x^2 + 3)] = \sec^2(2x^2 + 3) \cdot d(2x^2 + 3)$$

But

$$d(2x^2 + 3) = d(2x^2) + d(3) = 4x\, dx.$$

Therefore

$$d[\tan(2x^2 + 3)] = 4x \sec^2(2x^2 + 3)\, dx.$$

6. Differentiate $\frac{1}{4} \sec(2\theta^2)$.

Solution. By formula (R), $2\theta^2$ replacing θ, we have

$$d[\tfrac{1}{4} \sec(2\theta^2)] = \tfrac{1}{4} \cdot d[\sec(2\theta^2)] = \tfrac{1}{4} \sec(2\theta^2) \tan(2\theta^2) \cdot d(2\theta^2),$$

But

$$d(2\theta^2) = 4\theta\, d\theta.$$

Therefore

$$d(\tfrac{1}{4} \sec 2\theta^2) = \theta \sec 2\theta^2 \tan 2\theta^2\, d\theta.$$

7. Differentiate $(\sin x)^3$.

Solution. This being a power of a variable, we have

$$d(\sin x)^3 = 3(\sin x)^2 \cdot d(\sin x)$$

$$= 3(\sin x)^2 \cdot \cos x\, dx,$$

or since $(\sin x)^3$ is generally written $\sin^3 x$.

$$d(\sin^3 x) = 3 \sin^2 x \cos x\, dx.$$

8. If $y = \csc^2(\tfrac{1}{2}x)$ find dy/dx.

Solution. By formula (S), $\csc^2$ being a power, we have

$$dy = d\left[\csc^2\left(\frac{x}{2}\right)\right] = 2 \csc\left(\frac{x}{2}\right) \cdot d\left[\csc\left(\frac{x}{2}\right)\right]$$

and

$$d\left[\csc\left(\frac{x}{2}\right)\right] = -\csc\left(\frac{x}{2}\right)\cot\left(\frac{x}{2}\right)\cdot d\left(\frac{x}{2}\right)$$

$$= -\frac{1}{2}\csc\left(\frac{x}{2}\right)\cot\left(\frac{x}{2}\right)dx$$

$$\therefore\quad dy = 2\csc\left(\frac{x}{2}\right)\left[-\frac{1}{2}\csc\left(\frac{x}{2}\right)\cot\left(\frac{x}{2}\right)dx\right]$$

$$= -\csc^2\left(\frac{x}{2}\right)\cot\left(\frac{x}{2}\right)dx.$$

$$\therefore\quad \frac{dy}{dx} = -\csc^2\frac{x}{2}\cot\frac{x}{2}.$$

9. $y = \cot a\theta$. Find dy/dt.

Solution. By formula (Q), a being constant,

$$dy = d[\cot (a\theta)] = -\csc^2 (a\theta)\cdot d(a\theta)$$

$$= -a\csc^2 a\theta\cdot d\theta$$

$$\therefore\quad \frac{dy}{dt} = -a\csc^2 a\theta\,\frac{d\theta}{dt}.$$

10. Differentiate $2\sqrt{\tan\theta}$.

Solution. Using formulas (G) and (P),

$$d(2\sqrt{\tan\theta}) = 2\cdot d(\sqrt{\tan\theta}) = 2\cdot\frac{d(\tan\theta)}{2\sqrt{\tan\theta}}$$

$$= \frac{\sec^2\theta\,d\theta}{\sqrt{\tan\theta}}.$$

28. Illustrative Problems. 1. The crank and connecting rod of a steam engine are three and ten feet long respectively, and the crank revolves at a uniform rate of 120 r.p.m. At what rate is the crosshead moving when the crank makes an angle of 45 degrees with the dead center line?

Solution. In Fig. 10 let OC represent the dead-center line and the circle the path of the crank pin P. Then C will represent the crosshead, CP the connecting rod and OP the crank. As P moves steadily round the circle in the direction shown C moves back and forth at different rates along OC. If OC = x and angle POC = θ, then we are to find dx/dt when $d\theta/dt = 2$ rev. per sec. = 4π radians per sec. To do this we must therefore express x as a function of θ.

In the figure let a = crank length = 3 feet and b = length of connecting rod = 10 feet. Draw PA perpendicular to OC. Then for varying positions of P, A and C will have different positions but always

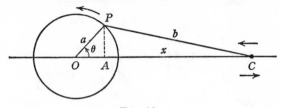

FIG. 10.

$$x = \overline{OA} + \overline{AC}. \tag{a}$$

In the right triangle PAC the hypotenuse formula gives

$$\overline{AC} = \sqrt{b^2 - \overline{AP^2}} \tag{b}$$

and in the right triangle POA by trigonometry

$$\overline{OA} = a \cos \theta, \quad \overline{AP} = a \sin \theta. \tag{c}$$

Substituting this value of $\overline{AP}$ in (b),

$$\overline{AC} = \sqrt{b^2 - a^2 \sin^2 \theta},$$

and this value of $\overline{AC}$ together with the value of $\overline{OA}$ in the first of equations (c), when used in equation (a) gives finally

$$x = a \cos \theta + \sqrt{b^2 - a^2 \sin^2 \theta}, \tag{d}$$

which expresses x as a function of the angle θ. In order to find the rate dx/dt this equation must be differentiated to get dx.

Differentiating equation (d) and carrying out the transformations and simplifications, this gives:

$$dx = d(a \cos \theta) + d(\sqrt{b^2 - a^2 \sin^2 \theta})$$

$$= a \cdot d(\cos \theta) + \frac{d(b^2 - a^2 \sin^2 \theta)}{2\sqrt{b^2 - a^2 \sin^2 \theta}}$$

$$= a(-\sin \theta \, d\theta) + \frac{-d(a^2 \sin^2 \theta)}{2\sqrt{b^2 - a^2 \sin^2 \theta}}$$

$$= -a \sin \theta \, d\theta - \frac{a^2(2 \sin^{2-1} \theta \cdot \cos \theta \, d\theta)}{2\sqrt{b^2 - a^2 \sin^2 \theta}}$$

$$= -a \sin \theta \, d\theta - \frac{a^2 \sin \theta \cos \theta \, d\theta}{\sqrt{b^2 - a^2 \sin^2 \theta}}$$

$$= -a \sin \theta \, d\theta - \frac{a^2 \sin \theta \cos \theta \, d\theta}{a \sqrt{\dfrac{b^2}{a^2} - \sin^2 \theta}}.$$

$$\therefore \quad dx = -a \sin \theta \left[1 + \frac{\cos \theta}{\sqrt{\left(\dfrac{b}{a}\right)^2 - \sin^2 \theta}} \right] d\theta.$$

$$\therefore \quad \frac{dx}{dt} = -a \sin \theta \left[1 + \frac{\cos \theta}{\sqrt{\left(\dfrac{b}{a}\right)^2 - \sin^2 \theta}} \right] \frac{d\theta}{dt}.$$

Now $a = 3$, $b = 10$, $\left(\dfrac{b}{a}\right)^2 = 11.1$, $\dfrac{d\theta}{dt} = 4\pi$.

$$\therefore \quad \frac{dx}{dt} = -12\pi \left(1 + \frac{\cos \theta}{\sqrt{11.1 - \sin^2 \theta}} \right) \sin \theta. \tag{e}$$

When $\theta = 45°$, $\sin \theta = .707$, $\cos 45° = .707$.

$$\therefore \quad \frac{dx}{dt} = -12\pi \left(1 + \frac{.707}{\sqrt{11.1 - (.707)^2}} \right) \times .707$$

$$= -32.44 \text{ ft./sec.}$$

Similarly when $\theta = 270°$, $\sin \theta = -1$, $\cos \theta = 0$.

$$\therefore \quad \frac{dx}{dt} = -12\pi \left(1 + \frac{0}{\sqrt{11.1 - (1)^2}} \right) \times (-1) = (-12\pi) \times (-1)$$

$$= +37.70 \text{ ft./sec.}$$

In the same way as here worked out for $\theta = 45°$ and $270°$ the velocity of the crosshead C, dx/dt, could be calculated for any value of θ, that is, any position of the crank OP. In the first case here worked out the negative value of dx/dt means that x is decreasing, that is, C is approaching O, the crosshead is moving toward the left. In the second case the positive sign means that it is moving to the right, x is increasing. In the same way formula (e) would give a negative value for any position of the crank pin P above the horizontal line OC and a positive value for any position below this line. This serves as a check and test of the differential formula, for it is at once seen by examining the figure that this is correct.

2. A man walks across a diameter, 200 feet, of a circular courtyard at the rate of five feet per second. A lamp on the wall at one end of a diameter perpendicular to his path casts his shadow on the circular wall. How fast is the shadow moving (i) when he is at the center, (ii) when 20 feet from the center, (iii) and at the circumference?

Solution.   In Fig. 11 let CB be the path of the man, and L the position of the lamp.   Let M represent the position of the man at any particular instant and y his distance from the center O.   Then P is the position of the shadow on the circular wall and s its distance AP along the curved wall from A, ds being the differential of s in the momentary direction of the tangent, which is continually changing so as to follow the curve, as described in article 2.

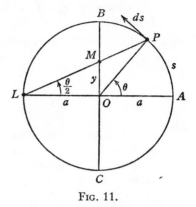

Fig. 11.

In order to find the rate of the shadow ds/dt, when that of the man dy/dt, is known we must as usual find the relation between y and s.

Draw OP and, as in Fig. 9, let angle AOP = θ, arc AP = s.   Also let the radius OL = OA = a; ds is indicated by the arrow.   If a line were drawn from the end of this arrow to L, the distance between the point where it crosses OB and the point M would be dy, the change in y corresponding to ds; and if a line were drawn from O to the end of the arrow, the angle between this line and OP would be $d\theta$, the change in θ corresponding to ds.

Using the notation just stated, we have as in Fig. 9

$$s = a \cdot \theta. \tag{a}$$

Also, by geometry, the angle PLA = $\frac{1}{2}$(angle POA).   Hence, angle MLO = $\frac{1}{2}\theta$ and in the right triangle MOL

$$y = a \tan\left(\tfrac{1}{2}\theta\right). \tag{b}$$

Differentiating (b) and (a),

$$dy = a \cdot d[\tan\left(\tfrac{1}{2}\theta\right)] = a \sec^2\left(\tfrac{1}{2}\theta\right) \cdot d(\tfrac{1}{2}\theta)$$

or,

$$dy = \tfrac{1}{2}a \sec^2\left(\tfrac{1}{2}\theta\right) d\theta; \quad ds = a\, d\theta.$$

From the first of these results, $d\theta = \dfrac{2}{a} \cdot \dfrac{1}{\sec^2 \frac{1}{2}\theta}\, dy$, and this value of $d\theta$ in the second gives

$$ds = \frac{2}{\sec^2 \frac{1}{2}\theta}\, dy. \qquad (c)$$

Since y and a are known at any time, it is convenient to use $\tan \frac{1}{2}\theta$ rather than the secant. In order to make the transformation we use the relation from trigonometry, $\sec^2 = 1 + \tan^2$ for any angle. Therefore, equation (c) can be written as

$$ds = \frac{2}{1 + \tan^2 \frac{1}{2}\theta}\, dy.$$

$$\therefore \quad \frac{ds}{dt} = \frac{2}{1 + \tan^2 \frac{1}{2}\theta}\, \frac{dy}{dt}. \qquad (d)$$

This gives the rate of the shadow in terms of the angle MLO and the rate of the man. Now, from the figure $\tan \frac{1}{2}\theta = y/a$ and we have given $a = 100$. Therefore, $\tan \frac{1}{2}\theta = y/100$, also $dy/dt = 5$, the rate of the man. Using these values, formula (d) becomes

$$\frac{ds}{dt} = \frac{10}{1 + \left(\dfrac{y}{100}\right)^2}.$$

When the position of the man is stated, his distance y from the center is known, and therefore the rate of the shadow on the wall ds/dt, is immediately calculated from this formula. We have three positions of the man given.

(i) When he is at the center $y = 0$, hence $ds/dt = 10$ ft./sec.

(ii) When he is 20 feet from the center $y = 20$, then

$y/100 = \frac{1}{5}$, hence $ds/dt = 10/[1 + (\frac{1}{5})^2] = 9\frac{8}{13}$ ft./sec.

(iii) When he is at the circumference $y = 100$, then

$y/100 = 1$ and $ds/dt = \frac{10}{2} = 5$ ft./sec.

This is a very interesting application of the trigonometric differentiation formulas, as it also includes the method of circular measure of angles and arcs and a very instructive trigonometric formulation and transformation, beside the differentiation of the tangent of an angle.

3. An elliptical cam arranged as in Fig. 12 rotates about the focus F as an axis and causes the roller at P to move up and down along the

line FP. If the diameters of the cam are six and ten inches and it rotates at the rate of 240 r.p.m., how fast is the roller moving at the moment when the long axis of the cam makes an angle of sixty degrees with the line of motion of the roller?

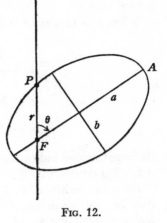

Solution. Let FP $= r$ and angle AFP $= \theta$, and let the half of the long and short diameters respectively be a and b. We then have given $d\theta/dt = 4$ rev./sec. $= 8\pi$ radians/sec., $a = 5$ in., $b = 3$ in., and are to find dr/dt when $\theta = 60°$. First we must have a relation between r and θ in order to find dr in terms of θ and $d\theta$. For the figure of an ellipse this is known to be

$$r = \frac{b^2}{a(1 - e \cos \theta)} \qquad \text{(a)}$$

where

$$e = \frac{\sqrt{a^2 - b^2}}{a}. \qquad \text{(b)}$$

Fig. 12.

In equation (a) therefore b and e are constants and we have to find dr by differentiating (a). Writing (a) as

$$r = \frac{b^2}{a} \cdot \left(\frac{1}{1 - e \cos \theta} \right),$$

$$dr = \frac{b^2}{a} \cdot d \left(\frac{1}{1 - e \cos \theta} \right)$$

and by the formula for a reciprocal,

$$dr = \frac{b^2}{a} \left[-\frac{d(1 - e \cos \theta)}{(1 - e \cos \theta)^2} \right] = \frac{b^2}{a} \left[\frac{d(e \cos \theta)}{(1 - e \cos \theta)^2} \right]$$

$$= \frac{b^2}{a} \left[\frac{-e \sin \theta \, d\theta}{(1 - e \cos \theta)^2} \right]$$

$$\frac{dr}{dt} = -\frac{b^2 e \sin \theta}{a(1 - e \cos \theta)^2} \frac{d\theta}{dt}. \qquad \text{(c)}$$

Since $a = 5$, $b = 3$, equation (b) gives $e = \frac{4}{5}$; also $b^2 = 9$ and $d\theta/dt = 8\pi$. Using these values, the cam data, (c) becomes

$$\frac{dr}{dt} = -\frac{57.6\pi \sin \theta}{5(1 - \frac{4}{5} \cos \theta)^2}.$$

This formula gives the instantaneous rate at which the roller is moving for any particular value of the angle between its line of motion and the long axis of the cam, and as the cam rotates about F the angle θ of course varies.

When $\theta = 60°$, $\sin \theta = \frac{1}{2}\sqrt{3}$ and $\cos \theta = \frac{1}{2}$. At this instant

$$\frac{dr}{dt} = -\frac{57.6\pi \times \dfrac{\sqrt{3}}{2}}{5(1 - \frac{4}{5}\cdot\frac{1}{2})^2} = -16\pi\sqrt{3}$$

$$= -87.1 \text{ in./sec.}$$

$$\frac{dr}{dt} = -7.25 \text{ ft./sec.,}$$

the negative sign indicating that r is decreasing, and therefore the roller is moving downward.

For other values of θ, that is, at other particular instants, the rate may be positive or negative, depending on the signs of the sine and cosine of the angle, for of course the roller will move both upward and downward in turn.

Exercises and Problems.

Find the differential of each of the following expressions:

1. $\sin 2x + 2 \sin x$.
2. $\cos^2 x + \sin^3 x$.
3. $\cos^3 x + \cos x^3$.
4. $\cos \sqrt{1 - t}$.

5. $\sin x \cdot \cos x$.
6. $x \sin x$.
7. $\frac{1}{2}\theta \cos 2\theta$.
8. $\frac{1}{4}(\tan 2x \cot 2x)$.

Find the derivative of each of the following expressions:

9. $\dfrac{\cos x}{x}$.
10. $\sin x \cdot \sin 2x$.
11. $\sqrt{\sec 2x}$.
12. $\frac{1}{3} \tan^3 x - \tan x + x$.
13. $\dfrac{1 + \cos x}{1 - \cos x}$.
14. $\sin (x + a) \cdot \cos (x - a)$, a constant.

Formulate each of the following problems and solve by the use of rates or derivatives as in article 28:

15. A person is approaching a 500-foot tower on a trolley car at the rate of ten miles per hour and looking at the top of the tower. At what rate must he be raising his head (or line of sight) when the car is 500 feet from the tower on level ground?

(*Hint:* Draw figure, call base line x, angle between x and line of sight θ, and find $d\theta/dt$ in radians and degrees per second, x decreasing.)

16. A signal observation station sights on a balloon which is rising steadily in a vertical line a mile away. At the moment when the angle of elevation of the telescope is 30 degrees and increasing at the rate of $\frac{1}{4}$ radian per minute, how high is the balloon and how fast is it rising?

17. A high-speed motor torpedo boat is moving parallel to a straight shore line at 40 land miles per hour, $1\frac{1}{2}$ miles from the shore, and is followed by a search-light beam which is trained on the boat from a station half a mile back from the shore. At what rate in radians per minute must the beam turn in order to follow the boat just as the boat passes directly opposite the station, and also when it is half a mile farther along the shore past the station?

18. A fighter plane is travelling at 300 miles per hour in a horizontal straight line and passes an enemy plane travelling in a parallel line at the same level at 250 miles per hour in the same direction. A gunner in the bomber trains his machine gun on the enemy plane as soon as he comes in range and turns the gun to keep his sights on the plane as he passes while firing on it. If the courses of the two planes are 200 yards apart how rapidly must the machine gun be turned to follow the enemy plane just as they pass, and also half a minute afterwards?

19. If $f = 2 \sin \theta - \cos 2\theta$, is f an increasing or decreasing function of θ when $\theta = 45°$, and when $\theta = 150°$? What is the rate of increase in each case, if θ is increasing at the rate of $\frac{1}{2}$ radian per minute?

20. The turning effect of a ship's rudder is shown in the theory of naval engineering to be $T = k \cos \theta \cdot \sin^2 \theta$, where θ is the angle which the rudder makes with the keel line of the ship. When the rudder is turning at $\frac{1}{4}$ radian per minute, what is the rate at which T is changing, in terms of the constant k, at the moment when $\theta = 30$ degrees?

Chapter 6

VELOCITY, ACCELERATION AND DERIVATIVES

29. Rate, Derivative and Velocity. We have based our idea of a differential on that of a rate and defined the differential of a variable y as the quantity dy which when divided by the corresponding quantity for time dt, gives the rate of the variable dy/dt.

We have, also, in Chapter 2, found that when the variable y is a function of an independent variable x, $y = f(x)$, then dy/dx is a new function of x, $f'(x)$, called the derived function or the derivative of y with respect to x. By comparison with dy/dt it is seen at once that dy/dt, *the rate of y, is simply the derivative of y with respect to t.* If, therefore, we have any quantity expressed as a function of t, we can at once find its rate by finding the derivative of the function with respect to t, which in turn is simply the differential of the function divided by dt.

As an illustration, consider the example given in article 9, $s = 16t^2$, in which s is the space in feet covered by a freely falling body in t seconds after beginning to fall. Now, when

$$s = 16t^2 \tag{34}$$

$$ds = d(16t^2) = 16 \cdot d(t^2) = 16 \cdot 2t \, dt$$

$$\frac{ds}{dt} = 32t. \tag{35}$$

This is the derivative of s, when s is the function of t expressed by equation (34).

Now, ds/dt is the rate of s, that is, the space rate of the falling body, and this is nothing more than the speed v. That is,

$$v = \frac{ds}{dt}. \tag{36}$$

As a second example, consider a train which travels in such a way that the distance covered in feet in any particular time after the mo-

52

ment of starting is equal to 44 times the number of seconds which have elapsed since the start. This can be expressed by

$$s = 44t, \tag{37}$$

$$\frac{ds}{dt} = 44,$$

that is,

$$v = 44 \text{ ft./sec.} \tag{38}$$

The result is, of course, simply the mathematical expression of the statement first made concerning the time and distance, for distance = (speed) × (time) at all times *when the speed is constant*. These last equations, therefore, are simply a test or check of the rate and derivative definitions in a case which is completely known in advance. Let us examine a little further, however, the ideas here involved.

30. Acceleration and Derivatives. There is another consideration of importance in connection with the two examples just given. In the case of the train, the speed is constant. It is well known, however, that the speed or velocity of a falling body is not constant, but steadily *increases* as long as the body continues to fall. This change in velocity is called *acceleration*. We have just seen that if we can give in symbols a description of the motion in the course of time, that is, express the distance covered as a function of the elapsed time, then the calculus provides a simple way to express the *velocity* at any time. The question next suggests itself, "Is there any such way to express the *acceleration*?"

The velocity is the rate of change of the distance, the derivative of s with respect to t. Similarly, the acceleration, being the rate of change of the velocity, must be the derivative of v with respect to t, that is (using a for acceleration),

$$a = \frac{dv}{dt}, \tag{39}$$

and this is the answer to our question.

It is sometimes convenient to write the derivative $\frac{dy}{dt}$ in the form $\frac{d}{dt}(y)$ in the same way as we may write the differential dy or $d(y)$. Thus we might say, "the derivative of y with respect to t," and write it $\frac{dy}{dt}$, or, "the derivative with respect to t, of y" and write it $\frac{d}{dt}(y)$. This form of writing the derivative is very convenient in the case of involved expressions which may be easily inclosed in brackets but which are not conveniently written as the numerator of a fraction with, say, dt as denominator.

Using this method, we may write equation (39) as

$$a = \frac{d}{dt}(v).$$

But, by (36),

$$v = \frac{d}{dt}(s).$$

Therefore,

$$a = \frac{d}{dt}\frac{d}{dt}(s),$$

that is, the *derivative of the derivative*, taken with respect to the same variable. This is called the *second derivative* of s with respect to t and written $\left(\frac{d}{dt}\right)^2(s)$ or $\frac{d^2s}{dt^2}$. Thus (39) becomes

$$a = \frac{d^2s}{dt^2}. \tag{40}$$

The figure "2" used in this symbol is *not* an exponent and does not indicate the square of the derivative, but simply indicates that the operation of differentiation is performed twice in succession on the same variable or function. The *square* of the derivative is denoted in the usual algebraic manner by writing $\left(\frac{dy}{dt}\right)^2$.

To illustrate the meaning of equation (39) or (40) let us apply it to the case of the freely falling body. We found that the velocity at any moment t is given by

$$v = \frac{ds}{dt} = 32t \text{ ft./sec.}$$

The acceleration is, therefore,

$$a = \frac{dv}{dt} = \frac{d^2s}{dt^2} = \frac{d}{dt}\left(\frac{ds}{dt}\right) = \frac{d}{dt}(32t) = 32\frac{dt}{dt}.$$

$$\therefore \quad a = 32 \text{ (ft./sec.) per sec.}$$

This states that the acceleration is constant which means that in this case the velocity increases uniformly, steadily. This acceleration is due to gravitation and the value $a = 32$ (more accurately 32.16) is referred to as the acceleration of gravity and generally denoted by the letter g. Acceleration in general, of any other value or cause, is generally denoted by a.

Next let us apply the equation (39) or (40) to the case of the train considered above. Since in this case the velocity is constant it has no rate of change, that is, there is no acceleration. This is expressed by saying that $a = 0$ and if our method is correct we should get this result by differentiating the velocity equation of the train. This is equation (38) and since

$$v = 44$$

$$a = \frac{dv}{dt} = \frac{d}{dt} (44) = 0,$$

since 44 is a constant and its differential and therefore its derivative is zero. Thus the acceleration is zero as expected.

The notion of a derivative with respect to time (rate) as velocity and the second derivative as acceleration is thus seen to be entirely consistent with the physical relation of these terms and their definitions.

31. Second and Higher Derivatives of Functions. In defining and deriving the second derivative of s with respect to t we had s expressed as a function of t. Similarly we define and derive the second derivative of any dependent variable with respect to its corresponding independent variable. Thus, if y is a function of x, we write for the second derivative $\frac{d^2y}{dx^2}$, or sometimes for convenience of printing d^2y/dx^2, as we write dy/dx. By analogy with the notation $f'(x)$ used to denote dy/dx, the second derivative of the function $f(x)$ is also written $f''(x)$. Thus, when

$$y = f(x)$$

and

$$\frac{dy}{dx} = f'(x),$$

then

$$\frac{d^2y}{dx^2} = f''(x). \tag{41}$$

In order to distinguish it from the second derivative, dy/dx is called the *first derivative*.

Now the second derivative is the derivative of the first derivative.

Similarly, the derivative of the second derivative is called the *third derivative* and written

$$\frac{d^3y}{dx^3} = f'''(x);$$

also

$$\frac{d^4y}{dx^4} = f^{iv}(x), \text{ etc.}$$

Thus certain functions may have any number of derivatives and the nth derivative would be written

$$\frac{d^ny}{dx^n} = f^n(x)$$

when, as above, y and x are the dependent and independent variables.

Depending on the form of the original function some of the derivatives may be equal to zero. Thus, in the case of the train as formulated above, when the function $s = f(t)$ is given by equation (37) as $f(t) = 44t$, we found that $f'(t) = 44$ and $f''(t) = 0$. Also, all other derivatives of this function after the first are zero. In the case of the falling body equation (34) gives $f(t) = 16t^2$ and we found that $f'(t) = 32t$ and $f''(t) = 32$, hence $f'''(t) = 0$, since 32 is a constant and its differential is zero. All other derivatives of this function are also zero.

If a certain derivative of a function is zero, it is said that the function does not have that derivative. Thus, $f(t) = 44t$ has a first but no second derivative, and $f(t) = 16t^2$ has a first and second but no third or other derivatives.

In geometrical and technical applications, the first and second derivatives are of the greatest importance; higher derivatives are not so much used.

In the next chapter we shall give an interpretation of the first and second derivatives of any function and find a useful application for them in the solution of a very important and interesting kind of problem.

32. Illustrative Examples.

1. Find the second derivative of $\frac{1}{6}x^3 + \frac{1}{2}x^2 + x + 1$.

Let

$$y = \frac{1}{6}x^3 + \frac{1}{2}x^2 + x + 1.$$

Then

$$\frac{dy}{dx} = \frac{1}{2}x^2 + x + 1.$$

$$\therefore \quad \frac{d^2y}{dx^2} = \frac{d}{dx}\left(\frac{dy}{dx}\right) = x + 1.$$

2. If $y = \sin x$ show that the second derivative equals $-y$. Since

$$y = \sin x, \quad dy = d(\sin x) = \cos x \, dx.$$

$$\therefore \quad \frac{dy}{dx} = \cos x, \quad d\left(\frac{dy}{dx}\right) = d(\cos x) = -\sin x \, dx.$$

$$\therefore \quad \frac{d^2y}{dx^2} = \frac{d}{dx}\left(\frac{dy}{dx}\right) = -\sin x = -y.$$

3. Find $\dfrac{d^2(\tan \theta)}{d\theta^2}$.

Let

$$y = \tan \theta, \quad dy = d(\tan \theta) = \sec^2 \theta \, d\theta.$$

$$\therefore \quad \frac{dy}{d\theta} = \sec^2 \theta, \quad d\left(\frac{dy}{d\theta}\right) = d(\sec^2 \theta) = 2 \sec \theta \cdot d(\sec \theta)$$

$$= 2 \sec \theta \cdot \sec \theta \tan \theta \, d\theta.$$

$$\frac{d^2y}{d\theta^2} = \frac{d^2(\tan \theta)}{d\theta^2} = 2 \sec^2 \theta \tan \theta.$$

4. In the following expression s is distance in feet and t is time in seconds. Find the velocity (v) and acceleration (a) at the end of 3 seconds from the start of the motion ($t = 3$):

$$s = \tfrac{1}{4}t^4 - \tfrac{1}{2}t^2 + t + 4.$$

$$v = \frac{ds}{dt} = t^3 - t + 1,$$

$$a = \frac{d^2s}{dt^2} = 3t^2 - 1.$$

Substituting in the formulas for v and a the value $t = 3$, we get

$$v = (3^3) - 3 + 1 = 27 - 3 + 1 = 25 \text{ ft./sec.}$$

$$a = 3(3^2) - 1 = 27 - 1 = 26 \text{ ft./sec.}^2$$

5. Given $s = \sin^2 t - 3 \cos t$; find v and a when $t = 1.5$.

$$v = \frac{ds}{dt} = 2 \sin t \cdot \cos t + 3 \sin t$$

$$= \sin (2t) + 3 \sin t,$$

$$a = \frac{d^2s}{dt^2} = 2 \cos (2t) + 3 \cos t.$$

Substituting in these formulas $t = 1.5$, we get

$$v = \sin 3 + 3 \sin 1.5,$$

and

$$a = 2 \cos 3 + 3 \cos 1.5,$$

and here the angular quantity is in radians, not degrees. Using a table of functions of angles in radians, or converting the radians to degrees, we find that

$$v = 0.1415 + 3 \times 0.9975 = 3.134 \text{ ft./sec.}$$
$$a = 2 \times (-0.9900) + 3 \times 0.0707$$
$$= -1.9800 + 0.2121 = -1.768 \text{ ft./sec.}^2$$

Exercises.

Find $\frac{d^2y}{dx^2}$ in each of the following:

1. $y = \frac{1}{x}$.

2. $y = \cos 2x$.

3. $y = x^4 - 12x^3 + 48x^2 - 50$.

4. $y = \frac{1}{3}x^3 - 2 \sin x$.

5. $y = \cos x$. Show that $\frac{d^2y}{dx^2} = -y$.

6. $y = \frac{x^3}{1 - x}$.

7. $y = \sqrt{4 + x^2}$.

8. $y = x \sin x$.

9. $y = \tan x$.

10. $y = \frac{\cos x}{x}$.

Find the velocity and acceleration in each of the following at the given time t in seconds when the distance s is in feet:

11. $s = 2t^3 - 15t^2 + 36t + 10$, $t = 1$.

12. $s = -2t^3 + 15t^2 + 36t$, $t = 2$.

13. $s = t \sin^2 t$, $t = \frac{\pi}{4}$.

14. $s = t^3 + \frac{1}{t}$, $t = 2$.

15. $s = 2 \sin t + \cos 2t$, $t = 1$.

16. A ball is thrown vertically upward with an initial velocity of 160 feet per second. The height s reached in t seconds is given by the equation $s = 160t - 16t^2$

    a. Find the velocity at the end of 2 seconds and at the end of 7 seconds.

    b. Find the acceleration.

17. An arrow is shot vertically upward. Its height h in feet after t seconds is given by the formula $h = 128t - 16t^2$. After how many seconds will the arrow stop rising and start falling?

Chapter 7

INTERPRETATION OF FUNCTIONS
AND DERIVATIVES
BY MEANS OF GRAPHS

33. Graphs and Functions. The method of representing statistical data, results of experiments, or corresponding sets of values of any two related quantities by graphs is familiar to everyone in technical work and business and in the social sciences. If the reader has studied the ARITHMETIC of this series he is familiar, not only with the method of plotting graphs, but also with the method of analyzing them and determining some of their properties, and from these the relations and properties of the quantities represented by the graphs. The subject of graphs is of great importance in calculus and for use in this and later chapters we discuss here some of the principal ideas connected with graphs which we shall need.

Let us begin with an illustration. Suppose that the average temperature for each month of the year for a certain neighborhood is given by the table below, being expressed in degrees Fahrenheit. A

Month		Average Temperature	Month		Average Temperature
Number	Name		Number	Name	
1	January	20°	7	July	80°
2	February	22	8	August	85
3	March	28	9	September	75
4	April	35	10	October	60
5	May	45	11	November	35
6	June	65	12	December	25

study and comparison of such figures will give a fair idea of how the weather changes as the time passes, but the table does not convey a complete single picture of the temperature variation for the entire year.

59

The above corresponding set of months and temperatures is plotted as a graph in Fig. 13, in which the numbers of the months are laid off horizontally and at the point or distance corresponding to each month is erected a vertical line of a certain length, which represents the corre-

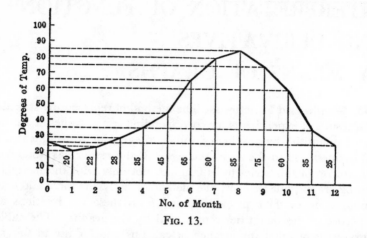

FIG. 13.

sponding number of degrees for that month as laid off on the vertical line at the left of the diagram.   The broken line joining the upper ends of these lines is called the *graph* of the average temperature for the year.

A glance at the graph shows the temperature *for any month* as well as the table.   It enables a comparison *for the different months* to be made at a glance, and it shows also in one glance the *variation* of the temperature as the time passes.   There is no relation between the temperature and the time of the year such that the temperature can be calculated or known exactly in advance, and as it varies slightly from year to year, it can only be determined by experience and observation.   The facts remain, however, that the *temperature depends on the time* of the year and that *to each time there corresponds a certain temperature*.

As a second example, suppose that a certain commodity is bought by weight at a set price per pound but delivered by truck at a certain price per trip without regard to weight, that is, up to the limit of the truck capacity.   The total cost will then be equal to the sum of the price, which depends on the number of pounds, and the delivery charge which is fixed.   The cost, therefore, varies with the weight. If the price per pound is 50 cts. ($\frac{1}{2}$) and the delivery charge is $5, then, if we represent the number of pounds by P and the total cost in dollars by C, we can represent it or calculate it by the formula

$$C = \tfrac{1}{2}P + 5. \qquad (42)$$

The total cost (C) is calculated by this formula for different numbers of pounds (P) and tabulated in the table below.

P	C	P	C
100	55	600	305
200	105	700	355
300	155	800	405
400	205	900	455
500	255	1000	505

These values are plotted in the graph of Fig. 14, in which the lines have lengths already marked on them and it is only necessary to mark a point to indicate the corresponding values of P and C.

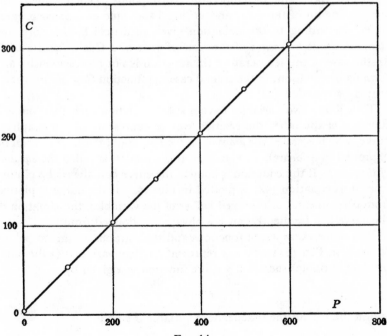

Fig. 14.

It is seen that the graph is a straight line and that for any value of P and not only for even hundreds, the corresponding value of C can be found by locating the value of P on the horizontal base line, from that point running up a vertical line to the graph, and then following the horizontal which there intersects the graph until the main vertical line at the left is reached. The corresponding value of C is then read on this line.

In each of the graphs, Figs. 13 and 14, the horizontal base line and the main vertical line are called the horizontal and vertical *axes*, the values read horizontally are called *abscissas*, the values read vertically are called *ordinates*, and the corresponding abscissa and ordinate for any point on the graph are called the *coordinates* of that point. In this connection the axes are also called the *coordinate axes*. The graph in Fig. 14 is called the *graph of the equation* (42) and (42) is called the *equation of the graph* of Fig. 14. In many cases graphs are not straight lines and the general name *curve* is used to apply to all graphs whether actually curved or straight.

In Fig. 13 a value of temperature corresponds to each value of the time or month of the year, and in Fig. 14 a value of C to each value of P. According to the definition given in article 8, therefore, the temperature is a *function* of the time of year and C is a *function* of P. In the case of the temperature the function is not represented by any formula or equation; in the second case the function $C = f(P)$ is represented by equation (42).

Thus if any two variables have a known relation such that one is a function of the other the relation can be represented by a graph or curve, which is called the *graph of the function*, and if the function can be expressed in a formula or equation, this equation is called the *equation of the curve*. If the equation is known, the curve can always be plotted from it as equation (42) is plotted in Fig. 14. If the curve is plotted from experimental or observed values of the variables, the equation of the curve can frequently but not always be obtained from it.

Two other examples of functions and their graphs are given below. In the first, Fig. 15, we let x, y represent any two variables (x the independent variable) such that y is the function of x given by

$$y = x^2, \tag{43}$$

and in the second the independent variable is θ and

$$y = \sin \theta. \tag{44}$$

The graph of this function is Fig. 16.

The graph of equation (43) is plotted by assigning values to x and calculating the corresponding values of y. The graph of (44) is plotted by assigning values in degrees to θ and finding the corresponding value of $\sin \theta$ from a table of natural sines.

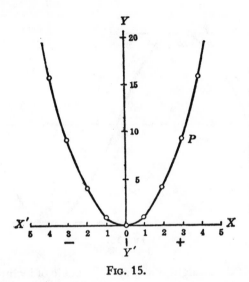

FIG. 15.

In Fig. 13 it is noted that in some parts of the curve the temperature increases as time passes and in others it decreases, rising to its highest value for no. 8 (August) and then falling off. The temperature is said to be an *increasing function* of the time up to no. 8 and after that point is passed, it is a *decreasing function*. Similarly, in Fig. 16 or equation (44), y is an increasing function of θ from $\theta = 0$ to $\theta = 90°$, from 90 to 270° y is a decreasing function of θ, and from 270 to 360° y is again an increasing function. In Fig. 14, or equation (42), however, and in Fig. 15 to the right of the y-axis (equation (43)), C and y always increase as P and x respectively increase. The *rate* of increase remains always the same for C in Fig. 14, but in the case of y in Fig. 15 the rate itself increases, and this rate of increase of y depends on x, y increasing more rapidly as x increases.

If the angle between the graph and the horizontal axis in Fig. 14 is indicated by ϕ, the quantity tan ϕ is called the *slope* of the graph. This is a common term in building (for roofs), for describing hill or embankment steepness (except in some kinds of railroad work), etc. That is, the slope is defined as the *tangent of the angle of inclination to the horizontal*.

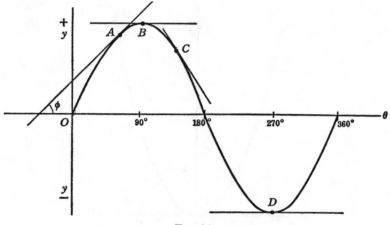

FIG. 16.

In the case of the straight line graph the angle of inclination and the slope both remain the same throughout its length and are easily pictured and determined. In the case of the curved lines, however, this is not the case. In the case of a curved graph such as Fig. 15, in order to determine the slope a tangent to the curve is drawn as at the point P (Fig. 15), and the angle ϕ is measured between this tangent line and the horizontal axis. Then, as before, tan ϕ is the slope of the curve. When the graph is not straight it is seen that the slope varies from point to point and we must therefore specify the particular point on the curve at which the slope is measured and have to refer to the *slope at a point*. Thus in Fig. 16 at A, the angle ϕ is less than 90° and the slope tan ϕ is positive, at C, ϕ is greater than 90° but less than 180° and the slope tan ϕ is negative. At B and D the tangent line is parallel to the horizontal axis and the slope is zero. At the points where the slope is *positive*, the tangent line is inclined to the right, and the function y *increases* as the independent variable increases. Where the slope is *negative*, the tangent line is inclined toward the left and the function is a *decreasing* function of the independent variable.

In particular, it is to be noted that at those points where the slope is zero the function is neither increasing nor decreasing but is just changing over from one to the other. Thus at B, Fig. 16, y has just ceased to increase and after B is passed it decreases. Thus, values of y are less on either side of B than at B and the point B is called a maximum point; it represents a maximum value of the function y, or simply a *maximum* of the function. Similarly values of y just on either side of D are greater (that is, points on the curve are higher) than at D, and D represents a *minimum* of the function y.

34. Differentials of Coordinates of a Curve. If we differentiate equations (42) and (43) we obtain

$$dC = \tfrac{1}{2}dP \qquad (42a); \qquad dy = 2x\,dx. \qquad (43a)$$

Since these differentials are found from the equations, it should be possible to represent them on the curves which are the graphs ("pictures") of the equations, and this is indeed the case.

Let Fig. 14, the graph of (42), be represented below by Fig. 14a.

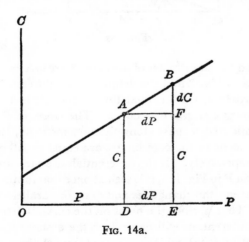

FIG. 14a.

Then the abscissa of any point A on the line is the distance OD = P, and the ordinate of A is DA = C. If we pass to a point B with a slightly greater abscissa OE, the difference between the two abscissas is DE, which is the P difference dP, and the difference between the ordinates is the C difference $FB = dC$.   That is, the differentials of the independent and dependent variables P, C in the equations (42), (42a) are represented by the changes in the coordinates which occur in passing from one point to another on the graph.

Next let us take the example of a graph which is not straight and represent the right hand half of Fig. 15, the graph of equation (43), by Fig. 15a below.   Here the tangent line PP′ makes the angle ϕ with

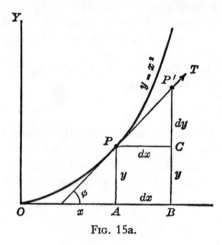

Fig. 15a.

the axis of x and the coordinates of the point P are $OA = x$ and $AP = y$. Referring back to Fig. 2 and the definition of a differential in article 3 and rate in article 2, we find that the differential is a change which occurs *in following a tangent* to a curve.   Therefore, in Fig. 15a as the point P is considered to pass along the tangent to P′, the changes dx and dy which occur in the coordinates are equal to AB and CP′ or to PC and CP′ respectively.   If the differentials are measured from some other point than P in Fig. 15a, it is seen at once that for the same change PP′ on the tangent, the changes dx and dy on or parallel to the axes will be different.   That is, since the point on the curve is determined by its abscissa, the differentials will depend on the abscissa x.   This is seen also from equation (43a), which states that the y changes, dy, will themselves change from point to point since dy depends on x, being equal to the change dx multiplied by $2x$.

In the case of the straight line graph the graph is its own tangent. In the case of the curved graph the differentials measured from a certain point on the curve are referred to the tangent to the graph at that point.   The differentials are in either case referred to the line whose inclination to the horizontal axis gives the slope of the curve at that point.   This is an important consideration, as will be seen in the next article.

35. Graphical Interpretation of the Derivative.   We have seen in article 29 that when the time t is the independent variable, the rate of any variable which is a function of t is simply the derivative of that variable with respect to t.   Based on the results of the last two articles we shall now find another interpretation of the derivative.

Let Fig. 17 represent the graph of any function y of the independent variable x and draw the tangent PT at the point P on the curve and

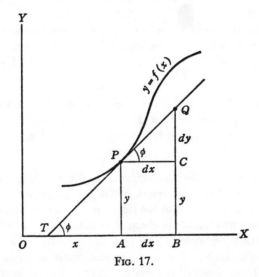

FIG. 17.

the coordinates $x = $ OA, $y = $ AP at the point P.   Then, in passing along the tangent to any neighboring point Q, the differentials of the coordinates are $dx = $ PC, $dy = $ CQ.   Also, the inclination of the tangent line at P is the angle BTP $= \phi = $ angle CPQ, and the slope of the curve is tan ϕ.   From trigonometry, in the right triangle CPQ,

$$\tan \phi = \frac{dy}{dx}. \tag{45}$$

But by definition (article 11) dy/dx is the derivative of the function y

with respect to x. We have, therefore, this important result: *If the graph of the equation representing any function*

$$y = f(x) \tag{46}$$

be plotted, then the derivative of y with respect to x,

$$\frac{dy}{dx} = f'(x), \tag{47}$$

is the slope of the curve.

As an illustration, consider the curve $y = x^2$, Figs. 15, 15a. Here

$$y = f(x) = x^2$$

and by (43a),

$$\frac{dy}{dx} = f'(x) = 2x.$$

This states that the slope of the curve at any point on the curve having the abscissa x is twice the abscissa, $2x$. For example, at the point $x = \frac{1}{2}$ the slope is $dy/dx = 1$, or since by (45) $\tan \phi = 1$,

$$\phi = 45°$$

is the angle of inclination of the tangent line to the horizontal. In the case of the straight line of Figs. 14 and 14a, equation (42a) gives for the slope $dC/dP = \frac{1}{2}$, that is, $\tan \phi = \frac{1}{2} = .500$. Hence,

$$\phi = 26\frac{1}{2}°.$$

In this case, the slope does not depend on the abscissa, and therefore, the angle ϕ is the same at every point on the graph.

The above interpretation of the derivative of a function with respect to the independent variable is of equal importance with the interpretation of the derivative with respect to time as a rate. In fact, we may consider the derivative of any dependent variable as its *relative rate with reference to the independent variable*. Thus, if dy/dt is the time rate of y, then dy/dx is its rate relative to x.

Now we have seen above in article 33 that when the slope of a curve at any point is positive, the function represented by the curve is increasing at that point as the independent variable increases, and decreasing when the slope is negative. Therefore, when the *derivative* is positive for any particular value of the independent variable the function is increasing at that point or for that value of the independent variable, and when the derivative is negative for any value of the independent variable, the function is decreasing. These relations are discussed in more detail and utilized in solving problems, in the next two chapters.

We give below some examples illustrating the application to curves of the principles developed in this chapter.

36. Illustrative Examples.

1. The equation of any straight line may be expressed in the form

$$y = mx + b. \tag{a}$$

What is the significance of each of the constants m, b?

Solution. If the equation (a) is differentiated,

$$dy = d(mx) - d(b) = m \, dx,$$

$$\therefore \quad \frac{dy}{dx} = m.$$

The constant m is, therefore, the slope of the line. If m is positive, the line is inclined toward the right; if negative, toward the left.

If in equation (a) we put $x = 0$, the resulting value of y gives the point where the line crosses the Y-axis, because at the Y-axis the abscissa is zero. When $x = 0$, we get

$$y = b.$$

The constant b is, therefore, the distance from the coordinate origin to the point where the line crosses the Y-axis. This is called *the y-intercept*. If b is positive, the line crosses the axis above the origin, if negative, below.

2. What is the nature of the graph of the equation

$$Ax + By + C = 0, \tag{a}$$

where A, B, C are constants?

Solution. By transposing the x-term and the constant term and dividing by B the equation becomes

$$y = -\frac{A}{B}x - \frac{C}{B}.$$

Comparing the equation in this form with (a) of example 1 we see at once that it represents a straight line and that the slope is

$$m = -(A/B)$$

and the y-intercept is

$$b = -(C/B). \tag{b}$$

These results do not mean that the slope or the y-intercept are negative, but that whether (A/B), (C/B) are positive or negative m and b are the negative of these quotients and may be either positive or negative.

To find the x-intercept, put $y = 0$ in equation (a). This gives $x_0 = -(C/A)$. The x-intercept is sometimes denoted by the letter a.

$$a = -(C/A). \tag{c}$$

By transposing the term C and dividing by $-C$ the equation (a) can be written in the form $\dfrac{x}{-\left(\dfrac{C}{A}\right)} + \dfrac{y}{-\left(\dfrac{C}{B}\right)} = 1$, and on comparing this with (b) and (c) it is seen at once that

$$\frac{x}{a} + \frac{y}{b} = 1$$

is another form of the equation of the straight line. This form is called the *intercept* form. The form (a) of example 1 is called the *slope-intercept* form and equation (a) above is called the *general* form of the equation of a straight line.

3. If the equation of a straight line is in the intercept form

$$\frac{x}{a} + \frac{y}{b} = 1, \tag{a}$$

find the relation between the differentials of the coordinates at any point (x, y) and also the slope of the line.

Solution. From the equation (a) we get by clearing fractions

$$bx + ay = ab,$$

and by differentiating this,

$$d(bx) + d(ay) = d(ab).$$

Therefore, since a, b are constants, $b \, dx + a \, dy = 0$. Hence,

$$a \, dy = -b \, dx \tag{b}$$

is the relation between the differentials.

Now divide equation (b) by the product $a \, dx$. This gives

$$\frac{dy}{dx} = -\frac{b}{a} \tag{c}$$

as the slope of the graph of (a).

4. A straight line graph is such that twice the differential of the

ordinate equals three times that of the abscissa.   Find an equation of the graph, the points where it crosses each axis, and the angle it makes with the X-axis.

Solution.   Formulating the given conditions,

$$2\,dy = 3\,dx.$$

Comparing this with equation (b) of example 3 we see at once that we can have

$$a = 2, \quad b = -3$$

as the x, y-intercepts.   That is, the graph can cut the X-axis at a point two units to the right of the origin and the Y-axis at a point three units below the origin.

Using these values of a, b in equation (a) of example 3, the equation of the graph is

$$\frac{x}{2} + \frac{y}{-3} = 1,$$

or,

$$3x - 2y = 6.$$

Another equation may also be found to meet the given conditions, but the graph would be parallel to the one represented by this one and the slope would be the same.   By transposing this equation it may be put into the general form or by solving it for y it may be put in the slope-intercept form, and in either of these forms the slope is easily determined as shown in the preceding examples.   By equation (c) in example 3, however, we find at once $dy/dx = \frac{3}{2}$.   That is,

$$m = 1.5$$

is the slope.   Since m is positive the line is, therefore, inclined to the right, and the angle of inclination to the X-axis is the angle ϕ such that $\tan \phi = m$.   Hence $\tan \phi = 1.5$ and by the tables $\phi = 56° 20'$.

5. Find the slope of the graph of the equation

$$x^2 - y + 2x = 6$$

at the point where the abscissa is $x = \frac{1}{2}$.

Solution.   Solving for y,

$$y = x^2 + 2x - 6.$$

Differentiating,

$$\frac{dy}{dx} = 2x + 2 = 2(x + 1)$$

is the slope at the point whose abscissa is x.   When $x = \frac{1}{2}$ it is

$$m = 2(\tfrac{1}{2} + 1) = 3.$$

6. What is the value of m at the point $(3, 4)$ on the graph of the equation $x^2 + y^2 = 25$?

Solution.   Differentiating the equation, $d(x^2) + d(y^2) = d(25)$, or,

$$2x\, dx + y\, dy = 0$$

$$\therefore \quad \frac{dy}{dx} = -\frac{x}{y}$$

and therefore, at the point where $x = 3$, $y = 4$, the slope is

$$m = -\tfrac{3}{4}.$$

7. The equation of a curve is $x^3 - 3x^2 - 3y + 6 = 0$.

(i) Find the inclination to the X-axis at the points where $x = 0$, and $x = 1$.

(ii) Find the points at which the tangent to the curve is parallel to the X-axis.

(iii) Find the point or points where the inclination of the tangent to the X-axis is 45°.

(iv) At what other point is the inclination the same as at $x = 3$?

Solution.   Solving the equation for y,

$$y = \frac{x^3}{3} - x^2 + 2.$$

Differentiating,

$$\frac{dy}{dx} = x^2 - 2x. \tag{a}$$

(i) Since the slope is $m = x^2 - 2x$, then at $x = 0$, $m = 0$ or $\tan \phi = 0$, hence $\phi = 0°$. At $x = 1$, $m = 1^2 - 2 = -1$, or $\tan \phi = -1$, hence $\phi = 135°$.

(ii) For the tangent line to be parallel to the X-axis the angle $\phi = 0$ and $m = \tan \phi = \tan 0° = 0$. Hence, by equation (a),

$$x^2 - 2x = 0.$$

Factoring this,

$$x(x - 2) = 0.$$

Hence,

$$x = 0, \quad \text{or} \quad x - 2 = 0,$$

and

$$x = 0, \quad x = 2$$

are the abscissas of the required points.

(iii) For the inclination $\phi = 45°$, $m = \tan \phi = 1$. Hence, by (a),

$$x^2 - 2x = 1,$$

or,

$$x^2 - 2x - 1 = 0,$$

is the condition which x must satisfy to make the tangent inclination 45°. To find the value or values of x satisfying this condition this quadratic equation is to be solved for x. Using the quadratic formula, we get

$$x = \frac{-(-2) \pm \sqrt{2^2 - 4(-1)}}{2} = \frac{2 \pm \sqrt{8}}{2} = \frac{2 \pm 2\sqrt{2}}{2} = 1 \pm \sqrt{2}.$$

$$\therefore \quad x = 1 + \sqrt{2} = 2.414$$

$$x = 1 - \sqrt{2} = -.414$$

are the abscissas at the required points.

(iv) At the point where $x = 3$ equation (a) gives as the slope

$$m = 3^2 - 2 \times 3 = 3.$$

To find the value of x at another point, where $m = 3$, put the slope

$$x^2 - 2x = 3.$$

Hence,

$$x^2 - 2x - 3 = 0, \quad \text{or} \quad (x - 3)(x + 1) = 0.$$

Therefore,

$$x - 3 = 0, \quad \text{or} \quad x + 1 = 0,$$

and

$$x = 3, \quad x = -1$$

are the abscissas at the points where the slope equals 3. The value $x = 3$ is the value first given. Therefore, the other required abscissa is $x = -1$.

This completes the solution. This example serves as a partial illustration of the manner in which the properties of a graph may be determined by means of the slope formula when its equation is known. In the case of curves plotted from geometrical data, this equation can always be obtained. In the case of a curve plotted from experimental or observed data, the equation can frequently be obtained exactly, and always approximately, and, therefore, its properties determined. From these properties, those of the quantities expressed by the equation can then be obtained. The equation derived from an experimentally plotted graph is called an *empirical equation* and in the sciences of physics, chemistry and engineering empirical equations are of the greatest importance.

8. For what values of θ is the tangent to the cosine curve horizontal?

Solution. The equation of the cosine curve is

$$y = \cos \theta.$$

Differentiating,

$$\frac{dy}{d\theta} = -\sin \theta$$

is the slope of the tangent line to the curve at any point given by the value of the angle θ. For the tangent to be horizontal the slope is zero. Therefore

$$-\sin \theta = 0, \quad \text{or} \quad \sin \theta = 0$$

and when the sine is zero we find from the tables that

$$\theta = 0° \quad \text{or} \quad \theta = 180°.$$

9. In passing along the graph of the equation

$$y = x^3 - 6x^2 + 3x + 5 \tag{a}$$

both the ordinate y and the slope m change, but generally at different rates. Find the point or points, if any, where the ordinate and slope momentarily change at the same rate.

Solution. The rate of change of the ordinate is dy/dt and of the slope, dm/dt. When these are the same, $dy/dt = dm/dt$.

Differentiating equation (a),

$$dy = (3x^2 - 12x + 3) \, dx,$$

and from this,

$$\frac{dy}{dt} = (3x^2 - 12x + 3) \frac{dx}{dt} \tag{b}$$

is the rate of change of the ordinate. Also from the same equation

$$\frac{dy}{dx} = m = 3x^2 - 12x + 3.$$

Differentiating this with respect to t to find the rate of m,

$$\frac{dm}{dt} = (6x - 12) \frac{dx}{dt} \tag{c}$$

is the rate of change of the slope. Placing $dy/dt = dm/dt$, we get from equations (b) and (c),

$$(3x^2 - 12x + 3) \frac{dx}{dt} = (6x - 12) \frac{dx}{dt},$$

or, cancelling the dx/dt on both sides of the equation,

$$3x^2 - 12x + 3 = 6x - 12.$$

This is the condition which must be fulfilled for the two rates to be equal. To determine the abscissa at the point or points where this condition holds, this equation must be solved for x. Transposing,

$$x^2 - 6x + 5 = 0.$$

Factoring,

$$(x - 5)(x - 1) = 0.$$

Hence,

$$x - 5 = 0, \quad \text{or} \quad x - 1 = 0,$$

and

$$x = 5, \quad x = 1$$

are the abscissas of the points at which the rates of change of the ordinate and the slope are equal.

Exercises.

Find the slope of the graph of each of the following equations at the point with the given abscissa or coordinates; also the angle the tangent makes with OX.

1. $y = x^4 + x^3 + x^2 + x + 1$; $x = -1$.
2. $3x + 4y - 6 = 0$; any point.
3. $y = 3x + 3 \sin x$; $x = \pi$.
4. $x^2 - y^2 = 16$; $x = 5, y = 3$.
5. $xy = 1$; $x = \frac{3}{2}$.
6. What is the value of x when the fraction $x^3/(x^2 + a^2)$ is changing at the same rate as x?

   (*Hint:* Let $y =$ the fraction; put $dy/dt = dx/dt$ and find x.)
7. Find the slope of the secant curve and the tangent curve at the point where each cuts the vertical axis.

   (*Suggestion:* Let $y = \sec x, y = \tan x$. Curves cut y axis when $x = 0$.)
8. A point moves in a circle whose equation is $x^2 + y^2 = 25$. At what two points on the circle is it moving momentarily parallel to the straight line $3x - 4y + 7 = 0$?
9. At what points on the curve $v = x^3 - 12x - 4$ does the slope equal 15?
10. Two particles remaining always in the same vertical line follow the

graphs of the equations $y = 2x^2 - 8x + 1$ and $2y = x^2 + 8x - 5$. How far from the Y axis will they be when moving momentarily parallel?

(*Hint:* The statement that they are in the same vertical line, ordinate line, means that x is the same for both.)

11. Given the equation $y = -x^2 + 6x - 1$. Write the equation of the line tangent to the graph of $y = -x^2 + 6x - 1$ at the point for which $x = 1$.

12. A tangent to the curve $y = x^2 - 4x + 10$ has a slope of 2. Find the coordinates of the point at which the tangent touches the curve.

Chapter 8

MAXIMUM AND MINIMUM VALUES

37. Maximum and Minimum Points on a Curve. In article 33 we have seen that corresponding to certain values of the independent variable a function may have values greater or less than those corresponding to other nearly equal values of the independent variable and that on the graph of the function these particular values are indicated by points which are higher or lower than neighboring points on either side. These points were there called *maximum* and *minimum* points respectively. Such points are shown at P and Q in Fig. 18. In this figure the irregular curve is taken as the graph of the equation.

$$y = f(x)$$

and $f(x)$ may represent any function whatever of the independent vari-

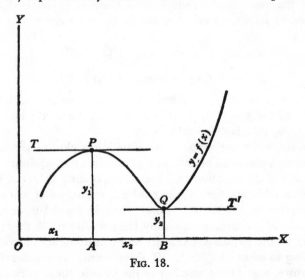

Fig. 18.

able x, that is, any form of expression involving x and constants only. Then corresponding to the abscissa $OA = x_1$ the value of the function is $AP = y_1 = f(x_1)$, this form of expression meaning that in the ex-

77

pression containing x, $f(x)$, the particular value x_1 is substituted. Similarly, corresponding to the abscissa OB $= x_2$ the value of the function is the ordinate BQ $= y_2 = f(x_2)$. We can express all this by saying that corresponding to the maximum point P, the maximum value of the function is y_1 and corresponding to the minimum point Q the minimum value of the function is y_2.

As seen in article 33, the tangent to the curve at a maximum or minimum point is parallel to the axis of abscissas and the slope of the curve is zero at those points. In Fig. 18 the tangent at P is PT and the tangent at Q is QT'. Now the slope of the graph of a function has been found to be the derivative of the function with respect to the independent variable. We have, therefore, the extremely important result: *When the value of a function is a maximum or a minimum value the derivative of the function is equal to zero.*

38. Maximum and Minimum Values of Functions. The result just obtained is expressed in symbols by saying that when $y = f(x)$ is a maximum or a minimum, then $f'(x) = 0$. In order to specify the point on the graph by giving the value of the abscissa of the point, or to name the value of the independent variable corresponding to the maximum or minimum value of the function we write that when (as in Fig. 18)

$$y_1 = f(x_1) = \text{max.}, \quad f'(x_1) = 0$$

or

$$y_2 = f(x_2) = \text{min.}, \quad f'(x_2) = 0$$

$$\left.\right\} \quad (48)$$

where the expressions $f(x_1)$, $f'(x_2)$, etc., mean that in the formulas or equations giving $y = f(x)$, $dy/dx = f'(x)$, etc., the particular value of x such as x_1, x_2, etc., is to be substituted.

Of course, when the graph of an equation is plotted, the maximum or minimum points can be seen at a glance and the maximum or minimum value of the function and the corresponding values of the independent variable are then easily read off, regardless of whether there is an equation for the curve or not. Thus, in Fig. 13 there is a maximum value of the average monthly temperature of 85 degrees corresponding to the month 8 (August) and a minimum value of 20 corresponding to month 1 (Jan.). If T represents the degrees of average temperature, and M the number of the month, then, although we do not know the form of the function, we know that T is a function of M and can write

$$T = f(M).$$

Then,
$$85 = f(8) = \text{max.}$$
and
$$20 = f(1) = \text{min.}$$

To state that at these points the derived function (derivative) is zero we can write
$$f'(8) = 0, \quad f'(1) = 0.$$

When the function is given, however, that is, the equation of the curve is known, it is not necessary to plot the curve for examination in order to locate the maximum or minimum values. It is only necessary to find the derivative $f'(x)$ of the equation $y = f(x)$ and put the derivative equal to zero. This gives a new equation $f'(x) = 0$ and the value of x to which the maximum or minimum value of y corresponds is found by solving this last equation for x. In order then to find the maximum or minimum value of y it is only necessary to substitute the value of x so found in the original equation and calculate y.

As an example consider the case of the function given by equation (43), $y = f(x) = x^2$. As in equation (43a) the derivative is $dy/dx = f'(x) = 2x$. The equation $f'(x) = 0$ then gives $2x = 0$ and hence $x = 0$ as the abscissa of the point at which the value of y is a minimum. This is verified by a glance at the curve Fig. 15.

As an example in which the curve is not plotted let us locate the maximum or minimum values of the function

$$y = x^2 - 4x + 5. \tag{49}$$

In this case

$$\frac{dy}{dx} = 2x - 4.$$

Hence for max. or min.,

$$2x - 4 = 0$$

$$x = 2.$$

In this case the value $x = 2$ gives a minimum value of y. To verify this, any values of x, positive or negative, may be substituted in the original equation (49) and y calculated for each. It will be found that for $x = 2$, y is less than for any other value of x. This minimum value of y is 1.

This example shows how we may not only find the value of x to which a maximum or minimum of y corresponds but also find that maximum or minimum value of y. Thus after finding that the minimum value of y corresponds to $x = 2$ we substitute this value of x in the original equation for y and calculate the corresponding value of y.

In general, therefore, in order to find the maximum or minimum value of a function substitute in it the value of the independent variable to which the maximum or minimum value corresponds, as found by placing the derived function (derivative) equal to zero.

39. Determining and Distinguishing Maxima and Minima. In the last article we saw how maximum or minimum values of a function may be located and determined either from the curve or the equation of the function. In discussing the examples given the question must have arisen in the mind of the reader as to how a maximum is to be distinguished from a minimum, or vice versa, without plotting the graph of the function. This question is easily answered with the means already at out disposal.

In article 33 we saw that when a curve rises to a maximum point as the abscissa increases and then falls off after the maximum point is passed, the slope changes from positive to negative. This is illustrated in Fig. 19. As x increases from the value x' at A to the value x'' at the maximum point P the slope of the curve changes from a positive

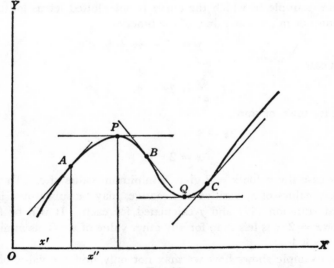

FIG. 19.

value to zero. As x increases beyond x'' the slope passes through the value zero and becomes negative. That is, *at P the slope decreases as x increases.* Similarly at B the slope is negative, at Q it is zero and there becomes positive as x increases, being positive again at C. That is, *at Q the slope increases as x increases.* These results may be expressed by saying that at maximum points the slope of a curve decreases and at minimum points increases as the abscissa increases.

When any variable increases, its rate is positive, and when it decreases, its rate is negative. Since the slope of a curve which has maximum or minimum points is a variable quantity, we have, therefore, the result that *at a maximum point the rate of the slope relative to the abscissa is negative and a minimum point it is positive.*

The slope is dy/dx when x, y are the coordinates and the rate of the slope relative to the abscissa x is $\dfrac{d}{dx}\left(\dfrac{dy}{dx}\right)$ or $\dfrac{d^2y}{dx^2}$. We can state finally, therefore, that *when a function $y = f(x)$ has a maximum the second derivative $f''(x)$ is negative and when it has a minimum the second derivative $f''(x)$ is positive.*

In order to determine whether $\dfrac{d^2y}{dx^2} = f''(x)$ is positive or negative, the function is differentiated twice in succession, and the value of x found by solving the equation $f'(x) = 0$ is substituted in the expression for the second derivative $f''(x)$. The value of $f''(x)$ is then calculated and will generally be either positive or negative, though certain special cases it may be neither but zero. Such cases will not be considered here.

As an example, let us consider the illustration used in the last article. We found that when $y = f(x) = x^2 - 4x + 5$, then $dy/dx = f'(x) = 2x - 4$ and when this is equal to zero $x = 2$. The second derivative is the derivative of $2x - 4$, that is, $f''(x) = 2$, which is positive. Since $f''(x)$ is positive then the value of y at $x = 2$ is a minimum, as was stated before.

Summarizing our results we can state the following rules for locating, distinguishing, and determining maximum and minimum values of a function:

   (i) Express the function in the form $y = f(x)$ by solving for y if not already so expressed.

   (ii) Find the first derivative $dy/dx = f'(x)$ and place this expression equal to zero.

(iii) Solve the equation so obtained for x. The value of x obtained will make y a maximum or a minimum.

(iv) Find the second derivative $f''(x)$ and substitute in this expression the value of x found in step (iii).

(v) Calculate the value of $f''(x)$ for this value of x. If it is positive y is a minimum and if it is negative y is a maximum.

(vi) To find the maximum or minimum value of y substitute the same value of x in the original expression and calculate y.

The application of these rules to a variety of functions is illustrated in the next article, and in the next chapter the method is applied to a number of interesting problems.

40. Illustrative Examples. 1. When is the function $y = (x - a)^3$ an increasing or decreasing function?

Solution. In order to answer this question we must find the conditions under which dy/dx is positive or negative. Differentiating $(x - a)^3$ as a power and dividing by dx we find

$$\frac{dy}{dx} = 3(x - a)^2.$$

If in this expression x is greater than a, $x - a$ is positive and so $(x - a)^2$ is positive. If x is less than a, $x - a$ is negative but the square is positive so that dy/dx is always positive and therefore the function y is always an increasing function.

2. Is $y = 1/x^3$ an increasing or decreasing function?

Solution. Differentiating, $dy/dx = -3/x^4$, and for any value of x, positive or negative, x^4 and therefore $1/x^4$ is positive, so that $-3(1/x^4)$ is negative. Since dy/dx is always negative the function y is a decreasing function.

3. Is $(3x + 5)/(x + 1)$ an increasing or decreasing function of the variable x?

Solution. Let $y = \dfrac{3x + 5}{x + 1}$. Differentiating the fraction, simplifying the result and dividing by dx, the derivative is

$$\frac{dy}{dx} = -\frac{2}{(x + 1)^2}.$$

This expression analyzed as above shows that for any value of x dy/dx is negative and therefore y is a decreasing function of x.

4. In example 7, article 36, for what values of x is y an increasing or decreasing function?

Solution. As given by equation (a) in that example

$$dy/dx = x(x - 2).$$

Any negative value of x makes each factor in this expression negative and therefore the product positive. Hence for any negative value of x, dy/dx is positive and y is an increasing function.

For any positive value of x less than 2, x is positive and $x - 2$ negative and therefore $x(x - 2)$ is negative and y a decreasing function; for any positive x greater than 2, x is positive and $x - 2$ also positive so that $x(x - 2)$ is positive and hence y increases.

Summarizing therefore, y is an increasing function for all values of x except zero to 2.

5. What value of x will give the expression $\frac{1}{2}x^2 - 3x$ its greatest or least value?

Solution. Let $y = \frac{1}{2}x^2 - 3x$; then the derivative is

$$\frac{dy}{dx} = x - 3$$

and for a maximum or minimum dy/dx must be zero. Hence

$$x - 3 = 0 \quad \text{and} \quad x = 3$$

gives y its greatest or least value.

6. Find the maximum value of the expression $32x - x^4$.

Solution. Let $y = 32x - x^4$. Then the derivative is

$$\frac{dy}{dx} = 32 - 4x^3.$$

Hence for maximum,

$$32 - 4x^3 = 0, \quad \text{or} \quad x^3 = 8.$$

$$\therefore \quad x = 2$$

is the value of x which makes y a maximum. Substituting this value of x in the original expression for y, we get

$$y_{\text{max.}} = 32 \cdot 2 - 2^4 = 64 - 16 = 48.$$

7. Show that the greatest value of the expression $\sin \theta + \cos 2\theta$ is $\frac{9}{8}$.

Solution. Let $y = \sin \theta + \cos 2\theta$. The derivative is then

$$\frac{dy}{d\theta} = \cos \theta - 2 \sin 2\theta.$$

Hence for maximum,

$$\cos \theta - 2 \sin 2\theta = 0.$$

Now by trigonometry $\sin 2\theta = 2 \sin \theta \cdot \cos \theta$. Using this in the last equation,

$$\cos \theta - 4 \sin \theta \cdot \cos \theta = 0.$$

Factoring, $\cos \theta (1 - 4 \sin \theta) = 0$, and hence $\cos \theta = 0$ or

$$1 - 4 \sin \theta = 0, \text{ and}$$

$$\therefore \quad \sin \theta = \tfrac{1}{4}$$

are the values of $\cos \theta$ and $\sin \theta$ which makes y a maximum. From this we must next find the value of $\cos 2\theta$ and then substitute in the expression for y to find its maximum value.

From trigonometry, $\cos 2\theta = 1 - 2 \sin^2 \theta$. Therefore

$$y = \sin \theta + 1 - 2 \sin^2 \theta$$

and using the maximizing value of $\sin \theta$,

$$y_{\text{max.}} = \tfrac{1}{4} + 1 - 2(\tfrac{1}{4})^2 = \tfrac{1}{4} + 1 - \tfrac{1}{8}$$
$$= \tfrac{9}{8}.$$

8. Find the maximum value of $a \sin x + b \cos x$.

Solution. Let $y = a \sin x + b \cos x$. Then, since a and b are constants, the derivative is

$$\frac{dy}{dx} = a \cos x - b \sin x.$$

Hence for maximum, $a \cos x - b \sin x = 0$, or $b \sin x = a \cos x$. Dividing both sides of this equation by $b \cos x$ we get

$$\sin x / \cos x = a/b.$$

But

$$\sin x / \cos x = \tan x.$$

Therefore the condition for y maximum is

$$\tan x = \frac{a}{b}.$$

This is the right triangle relation when a is the altitude and b the base. Hence the hypotenuse is $c = \sqrt{a^2 + b^2}$. From this we get, since $\sin x = a/c$ and $\cos x = b/c$,

$$\sin x = \frac{a}{\sqrt{a^2 + b^2}}, \quad \cos x = \frac{b}{\sqrt{a^2 + b^2}}$$

as the final conditions for y to be a maximum. Substituting these values in the original expression for y, we find

$$y_{\text{max.}} = a\left(\frac{a}{\sqrt{a^2 + b^2}}\right) + b\left(\frac{b}{\sqrt{a^2 + b^2}}\right) = \frac{a^2 + b^2}{\sqrt{a^2 + b^2}} = \sqrt{a^2 + b^2}.$$

9. Find the position and length of the greatest and of the least ordinate of the graph of the equation

$$y = 3x^3 - 9x^2 - 27x + 30. \tag{a}$$

Solution.   Differentiating,

$$\frac{dy}{dx} = 9x^2 - 18x - 27. \tag{b}$$

Hence for maximum or minimum,

$$9x^2 - 18x - 27 = 0.$$

Dividing by 9 and factoring, $x^2 - 2x - 3 = (x - 3)(x + 1) = 0.$

$$\therefore \quad x = 3, \quad x = -1 \tag{c}$$

are the abscissas at which are located the maximum and minimum ordinates.   In order to distinguish between them we must find the second derivative $f''(x)$.   Differentiating equation (b),

$$f''(x) = 18x - 18. \tag{d}$$

Using in this the first value of x from (c) we find

$$f''(3) = 18 \cdot 3 - 18 = +36$$

and since this is positive the least (minimum) ordinate occurs at the point where $x = 3$.   Using the second value of x from (c) in (d) we get

$$f''(-1) = 18(-1) \cdot\!\!- 18 = -36$$

and this being negative the greatest (maximum) ordinate occurs at the point where $x = -1$.

In order to find the length of the greatest and least ordinates we substitute the values of the corresponding abscissas from (c) in the equation (a) giving y.   Thus we find at $x = -1$,

$$y_{\text{max.}} = 3(-1)^3 - 9(-1)^2 - 27(-1) + 30 = +45,$$

and at the point where $x = 3$,

$$y_{\text{min.}} = 3(3)^3 - 9(3)^2 - 27(3) + 30 = -51.$$

It is to be noted that in these results the word *greatest* does not necessarily mean actually *longest*, nor does *least* mean actually *shortest*. But mathematically all negative numbers are "less" than positive numbers, or graphically, the end of the ordinate $y = -51$ is below the end of the ordinate $y = +45$.

Exercises.

Find the values of x which make y a maximum or a minimum in each of the following and find the corresponding maximum and minimum values of y:

1. $y = 2x^3 - 3x^2 - 36x + 25$.

2. $3y = x^3 - 6x^2 - 36x + 30$.

3. $y = x^4 - 8x^3 + 22x^2 - 24x + 12$.

4. $y = \dfrac{6x}{x^2 + 1}$.

5. $y = x - \dfrac{108}{x^2}$.

6. $xy - x^2 + y = 0$.

7. $y = \sin x - 2 \cos x$.

8. $y = \sin x + \tan x$

9. $y = 2 \sin x - \sin 2x$.

10. $y = \frac{1}{4} \cos^2 x - \sin 2x$.

Chapter 9

PROBLEMS IN MAXIMA AND MINIMA

41. Introductory Remarks. In this chapter we take up the extremely important matter of the application of the maximum and minimum method of the calculus to actual problems of ordinary observation, technical problems, etc.

Before the calculus method can be used, the relation between the known variables and the quantity whose variation is to be examined must be known, that is, the quantity whose maximum or minimum is sought must be expressed as a function of the controlling variable. The derivative of the dependent variable with respect to the independent variable is then found and placed equal to zero, and the resulting equation solved by the ordinary methods of algebra. In other words, after the functional relation is once expressed the procedure given in article 39 is to be followed exactly in cases where a complete investigation of the relation is to be made.

In many cases the function can have only a maximum or only a minimum value and not both; in other cases, even when both may exist, the nature of the problem, or the form of the equation obtained when the first derivative is placed equal to zero, will determine whether the result is a maximum or a minimum. In such cases it is not necessary to find the second derivative. The different cases are illustrated and discussed in the following article.

42. Illustrative Problem Solutions. 1. Divide the number 10 into two parts such that the product of the square of one part and the cube of the other shall be the greatest possible.

Solution. The lettered steps in the solution of this problem will serve as a model in solving other problems in maxima and minima.

a. Identify and express the function which is to be a maximum.

Let x = the part of 10 that is to be cubed
And $(10 - x)$ = the part of 10 that is to be squared

We could have let $x = $ the part that is to be squared had we wished but the above formulation will yield a simpler solution.

Thus, the function, y, that is to be a maximum may be written as follows.

$$y = (10 - x)^2 \cdot x^3$$

b. Differentiate y with respect to x.   Use the product rule.

$$y = (10 - x)^2 \cdot x^3$$

$$dy = d[(10 - x)^2 \cdot x^3] = (10 - x)^2 \cdot d(x^3) + x^3 \cdot d[(10 - x)^2]$$

$$= (10 - x)^2 \cdot 3x^2 \, dx + x^3 \cdot 2(10 - x) \cdot d(10 - x)$$

$$= [3x^2(10 - x)^2 - 2x^3(10 - x)] \, dx$$

$$\frac{dy}{dx} = 3x^2(10 - x)^2 - 2x^3(10 - x)$$

Take out the factor $x^3(10 - x)$ to obtain the result

$$\frac{dy}{dx} = x^3(10 - x)[3(10 - x) - 2] = x^3(10 - x)[30 - 3x - 2x]$$

$$= x^3(10 - x)(30 - 5x) = 5x^3(10 - x)(6 - x)$$

c. Set $\frac{dy}{dx} = 0$ and solve for x

$$5x^3(10 - x)(6 - x) = 0$$

Since this product is zero, any one of the factors may be zero.

$$5x^2 = 0, \quad 10 - x = 0, \quad 6 - x = 0$$

$$x = 0, \quad\quad x = 10, \quad\quad x = 6$$

d. Examine the above values for a maximum

If $x = 0$, $10 - x = 10$, and the product $(10 - x)^2 \cdot x^3 = 0$.   This is obviously a minimum value.

If $x = 10$, $10 - x = 0$, and again the product $(10 - x)^2 \cdot x^3 = 0$. Again, we have a minimum value.

If $x = 6$, $10 - x = 4$, and the product $(10 - x)^2 \cdot x^3 = (4)^2(6)^3 = 3456$.

This is a maximum value.   We may verify this result by finding the second derivative, substituting the value $x = 6$ in the second derivative, and discovering that the second derivative is negative.

In this case, it is not necessary to use the second derivative case since there is no question that $x = 6$ yields a maximum value.

In this example, we have not only found that the division of 10 into 6 and 4 will yield a maximum value but we have also discovered that the maximum value of the function is 3456.

2. A piece of cardboard one foot square has a small square cut out at each corner in order to turn up the edges of the remainder to form a box with a square base. Find the side of the small square cut out in order that the box may have the greatest possible capacity.

Solution. Let x represent the side of the small square. The form and dimensions of the cut piece are then shown in Fig. 20, the card-

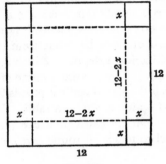

FIG. 20.

board being cut along the solid lines and folded along the dotted lines, and the dimensions being in inches. From this figure we are to formulate the volume V in terms of x and the known dimensions and find the value of x which makes V a maximum.

From the figure the side of the square base of the box is $12 - 2x$ or, factoring, $2(6 - x)$ and the depth of the box is x. The volume is the product of the base area by the depth, that is $[2(6 - x)]^2 \cdot x$. Hence

$$V = 4x(6 - x)^2. \tag{a}$$

This expresses V as a function of x and we are to find x for $V_{max.}$:

Differentiating, and noting that the 4 is a constant coefficient,

$$dV = 4 \cdot d[x \cdot (6 - x)^2] = 4\{x \cdot d[(6 - x)^2] + (6 - x)^2 \cdot d(x)\}$$
$$= 4\{x \cdot 2(6 - x) \cdot d(6 - x) + (6 - x)^2 \cdot dx\}$$
$$= 4\{2x(6 - x) \cdot (-dx) + (6 - x)^2 \, dx\}$$
$$= 4\{-2x(6 - x) + (6 - x)^2\} \, dx.$$

$$\therefore \quad \frac{dV}{dx} = 4[(6 - x)^2 - 2x(6 - x)].$$

Simplifying this by factoring it becomes

$$\frac{dV}{dx} = 12(6 - x)(2 - x).$$

For V to be a maximum dV/dx must be zero; hence

$$12(6 - x)(2 - x) = 0.$$

$$\therefore \quad 6 - x = 0, \quad \text{or} \quad 2 - x = 0,$$

and

$$x = 6, \quad \text{or} \quad x = 2$$

are the values of x which make V a maximum or a minimum.

If $x = 6$ the square is simply cut twice across the center into four 6-inch squares and no box is made. Hence it must be that $x = 2$ and the box of maximum content to be made from the foot-square cardboard has a base of 8 inches, a depth of 2 inches, and by putting this value of x in equation (a), a maximum volume of 128 cubic inches.

3. A funnel of which the flared conical part is 9 inches deep and 12 inches across is placed upside down on a table. What are the dimensions of the largest cylindrical can which can just be placed upright underneath the funnel to touch all round but not hold the funnel off the table?

Solution. The arrangement is shown in Fig. 21. BC = 12 inches is

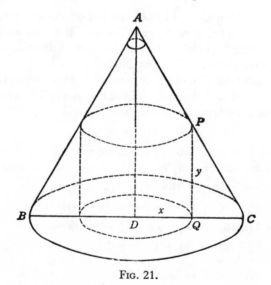

Fig. 21.

the diameter of the funnel flare and AD = 9 inches is its depth; the radius is DC = 6 inches. Then PQ is the depth of the can and DQ the radius of its base. The values of these two dimensions are to be found which make the volume of the can the greatest possible consistent with the requirement that it must just fit under the funnel cone.

Let DQ = x and PQ = y; the volume of the can is then

$$V = \pi x^2 y, \tag{a}$$

and the values of x and y are to be found which make V a maximum. There are two unknowns, however, x and y, so one of these must be expressed in terms of the other so that there will be only one independent variable in terms of which the dependent variable V is expressed.

Since the entire figure is symmetrical the triangles ADC, PQC are similar. Therefore, $\overline{PQ}:\overline{QC}::\overline{AD}:\overline{DC}$, or, since $\overline{PQ} = y$ and $\overline{QC} = \overline{DC} - \overline{DQ} = \overline{DC} - x$, the proportion becomes

$$\frac{y}{\overline{DC} - x} = \frac{\overline{AD}}{\overline{DC}}.$$

Putting in the values of AD and DC, we get $\dfrac{y}{6 - x} = \dfrac{9}{6}$, and hence,

$$y = \tfrac{3}{2}(6 - x). \tag{b}$$

Substituting this value of y in equation (a), we have for the volume,

$$V = \frac{3\pi}{2} x^2(6 - x), \tag{c}$$

and the value of x is to be determined which makes V a maximum. Differentiating,

$$dV = \frac{3\pi}{2} \cdot d[x^2 \cdot (6 - x)] = \frac{3\pi}{2} [x^2 \cdot d(6 - x) + (6 - x) \cdot d(x^2)]$$

$$= \frac{3\pi}{2} [x^2 \cdot (-dx) + (6 - x) \cdot 2x \, dx]$$

$$= \frac{3\pi}{2} [-x^2 + 2x(6 - x)] \, dx.$$

$$\therefore \quad \frac{dV}{dx} = \frac{3\pi}{2} [2x(6 - x) - x^2].$$

Simplifying this, it becomes

$$\frac{dV}{dx} = \frac{9\pi}{2} x(4 - x).$$

Hence for V maximum,

$$\frac{9\pi}{2} x(4 - x) = 0,$$

and

$$x = 0, \quad \text{or} \quad x - 4 = 0$$

are the solutions which make V either a maximum or a minimum. If $x = 0$ there is no cylinder, or rather the cylinder has zero volume, which is the minimum. Therefore we must use $x - 4 = 0$.

$$\therefore \quad x = 4 \tag{d}$$

is the radius of the can of maximum volume.

Putting this value of x in formula (b) we find that the depth is

$$y = 3.$$

From (d) the diameter of the can is 8. The largest can which will just go under the inverted funnel is therefore a flat can 3 inches deep and 8 inches across. The volume of this can is given by using $x = 4$ in equation or formula (c) or $x = 4, y = 3$ in (a) and is found to be

$$V = 150.8 \text{ cu. in.}$$

Incidentally, from the dimensions given for the conical funnel and the cylindrical can (counting the entire volume of the funnel cone, to the point A in Fig. 21), the volume of the funnel cone is 1357.2 cu. in. and therefore the volume of the can is $\frac{1}{9}$ that of the funnel top.

4. A man in a boat five miles from the nearest point on a straight beach wishes to reach in the shortest possible time a place five miles along the shore from that point. If he can row four miles an hour and run six miles an hour where should he land?

Solution. If the landing place is specified by stating its distance from the nearest point on the beach to the boat, then the total time required to reach the final destination is to be formulated in terms of this distance, and this distance determined such that the time is a minimum.

In Fig. 22 let OP be the beach line and B the position of the boat at a distance OB = 5 miles from the nearest point O on the beach; and let P be the place which the man wishes to reach, at a distance OP = 5

miles from O. He must then row straight to some point A on the
beach at the distance OA $= x$ from O and run the remaining distance
AP $= 5 - x$. We are to formulate the time taken to row the distance
BA and run the distance AP and find the value of x which makes this
time a minimum.

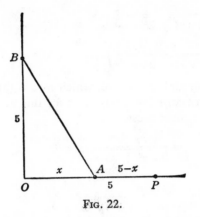

Fig. 22.

In the right triangle AOB the distance he must row is

$$\overline{BA} = \sqrt{\overline{BO}^2 + \overline{OA}^2} = \sqrt{x^2 + 25}$$

and at 4 miles an hour the time it will take him is $t_1 = \overline{BA}/4$. Therefore

$$t_1 = \tfrac{1}{4}\sqrt{x^2 + 25}. \tag{a}$$

After landing he must run the distance AP $= 5 - x$ and at 6 miles an
hour it will require a time $t_2 = \overline{AP}/6$. Therefore

$$t_2 = \tfrac{1}{6}(5 - x). \tag{b}$$

The total time required will be $t = t_1 + t_2$. Hence, adding (a), (b),

$$t = \tfrac{1}{4}\sqrt{x^2 + 25} + \tfrac{1}{6}(5 - x). \tag{c}$$

This is the required formulation of the time as a function of the distance x and the value of x is to be determined which makes t a minimum.

Differentiating equation (c),

$$dt = \tfrac{1}{4} \cdot d(\sqrt{x^2 + 25}) + \tfrac{1}{6} \cdot d(5 - x)$$

$$= \frac{1}{4} \cdot \frac{d(x^2 + 25)}{2\sqrt{x^2 + 25}} + \frac{1}{6} \cdot d(-x) = \frac{1}{4} \cdot \frac{2x\,dx}{2\sqrt{x^2 + 25}} - \frac{1}{6}\,dx$$

$$= \frac{1}{2}\left[\frac{x}{2\sqrt{x^2 + 25}} - \frac{1}{3}\right]dx.$$

$$\therefore \quad \frac{dt}{dx} = \frac{1}{2}\left(\frac{x}{2\sqrt{x^2 + 25}} - \frac{1}{3}\right)$$

For t to be a minimum, $dt/dx = 0$, which means that the product on the right of the last formula is zero, and for this to be zero, since the factor $\frac{1}{2}$ is not zero, then must

$$\frac{x}{2\sqrt{x^2 + 25}} - \frac{1}{3} = 0.$$

Transposing,

$$\frac{x}{\sqrt{x^2 + 25}} = \frac{2}{3}.$$

Squaring,

$$\frac{x^2}{x^2 + 25} = \frac{4}{9}.$$

Clearing fractions,

$$5x^2 = 100.$$

Therefore

$$x^2 = 20,$$

and

$$x = \sqrt{20} = \pm 4.47 \text{ miles.}$$

If we use the negative value the landing place must be 4.47 miles to the *left* of O in Fig. 22 which obviously will not allow the time to be a minimum. He must therefore land 4.47 miles from O in the direction of P, or a little more than half a mile from his destination.

5. What is the cheapest form of cylindrical tin can of given volume?

Solution. Let the diameter be D, the height H and the specified (constant) volume V. Then

$$V = \tfrac{1}{4}\pi D^2 H. \qquad\qquad (a)$$

The amount of sheet tin required to make the can is equal to the sum of the areas of the circular ends and the curved cylindrical surface. The end area is $\tfrac{1}{4}\pi D^2$ and for the two, $\tfrac{1}{2}\pi D^2$. The curved surface is the product of the circumference and height, that is, πDH. The total surface area is therefore $\frac{1}{2}\pi D^2 + \pi DH$, or if we let S represent the total surface,

$$S = \pi(\tfrac{1}{2}D^2 + DH). \tag{b}$$

The relation between D and H is to be determined so that with V constant as given in (a), S shall be a minimum.

In order to have only one independent variable in the surface formula (b) let us express H in terms of D by means of equation (a).   Solving (a) for H we get $H = \dfrac{4V}{\pi}\dfrac{1}{D^2}$, and substituting this value of H in (b) gives, when simplified,

$$S = \tfrac{1}{2}\pi D^2 + 4V\frac{1}{D},$$

which gives S as a function of D.

Differentiating this, π and V being constant,

$$dS = \frac{\pi}{2} \cdot d(D^2) + 4V \cdot d\left(\frac{1}{D}\right)$$

$$= \frac{\pi}{2} \cdot 2D\,dD + 4V \cdot \left(-\frac{dD}{D^2}\right) = \left(\pi D - \frac{4V}{D^2}\right)dD.$$

$$\frac{dS}{dD} = \pi D - \frac{4V}{D^2}.$$

For S to be a minimum, $dS/dD = 0$.   Therefore

$$\pi D - \frac{4V}{D^2} = 0$$

and

$$D^3 = \frac{4V}{\pi}. \tag{c}$$

Now by equation (a), $4V = \pi D^2 H$.   Substituting this in equation (c) to eliminate V we get as the condition for minimum surface,

$$D^3 = \frac{\pi D^2 H}{\pi}.$$

$$\therefore \quad D = H$$

is the relation between D and H for the minimum amount of tin plate. That is, for the cheapest can the height and diameter are the same.

6. An electric dry cell of electromotive force E volts and internal resistance r ohms is connected in series with an external resistance R

ohms. What value should R have in order that the greatest power be supplied to the circuit?

Solution. The total resistance of the circuit consisting of R and r ohms in series is $R + r$ ohms and therefore by Ohm's Law the current flowing is $E/(R + r)$. The power P in watts supplied to the external circuit of resistance R is the product of the current squared by the resistance R, or $\left(\dfrac{E}{R + r}\right)^2 \cdot R$. Hence

$$P = E^2 \frac{R}{(R + r)^2}.$$

In this expression E and r are constants and the value of R is to be determined which makes P a maximum.

Differentiating,

$$dP = E^2 \cdot d\left[\frac{R}{(R + r)^2}\right] = E^2 \left\{\frac{(R + r)^2 \cdot d(R) - R \cdot d[(R + r)^2]}{[(R + r)^2]^2}\right\}$$

$$= E^2 \left\{\frac{(R + r)^2 \, dR - R \cdot [2(R + r) \cdot d(R + r)]}{(R + r)^4}\right\}$$

$$= E^2 \left\{\frac{(R + r)^2 \, dR - 2R(R + r) \, dR}{(R + r)^4}\right\}$$

$$= E^2 \left\{\frac{(R + r)^2 - 2R(R + r)}{(R + r)^4}\right\} dR$$

$$= E^2 \left[\frac{(R + r)(r - R)}{(R + r)^4}\right] dR = E^2 \left[\frac{r - R}{(R + r)^3}\right] dR.$$

$$\therefore \quad \frac{dP}{dR} = E^2 \left[\frac{r - R}{(R + r)^3}\right].$$

For P to be a maximum, the product on the right must be equal to zero and since E^2 is not zero, then,

$$\frac{r - R}{(r + R)^3} = 0,$$

and for the fraction to be zero, its numerator must be zero.

$$\therefore \quad r - R = 0 \quad \text{and} \quad R = r$$

is the condition for maximum power in the external circuit. That is, the greatest power is supplied to the circuit when the external resistance of the circuit equals the internal resistance of the battery.

7. A produce house has in cold storage ten thousand pounds of beef

worth at present sixteen cents a pound wholesale. If the price increases steadily one cent per week while the beef loses a hundred
pounds a week in weight and the fixed overhead storage charges are
sixty dollars a week, how long should the beef be held before selling
for the greatest net value?

Solution.   Here we must formulate the net value at any number of
weeks n from the present time and find n such that the total net value
at that time shall be a maximum.

With 100 lb. a week shrinkage the total weight in n weeks will be
$10,000 - 100n$, or $100(100 - n)$. The present price is $16\cent = \frac{4}{25}$
dollars and with a $1\cent = \frac{1}{100}$ dollar advance per week the price per
pound in n weeks will be $(4/25) + (1/100)n = \left(4 + \dfrac{n}{4}\right)/25$. The
gross value will then be the weight times the price per pound, or
$100(100 - n) \cdot \left(4 + \dfrac{n}{4}\right)/25$. The gross value at the end of n weeks
is, therefore, multiplying this out and simplifying,

$$V_g = 1600 + 84n - n^2. \tag{a}$$

In the meantime, however, there is a steady charge of 60 dollars a
week which in n weeks amounts to $60n$ dollars.   Subtracting this from
the gross value given by equation (a) the net value at the end of n
weeks is $V_g - 60n = V$.   The net value is therefore

$$V = 1600 + 24n - n^2,$$

and the value of n is to be found which makes V a maximum.

Differentiating,

$$dV = 24\ dn - 2n\ dn = 2(12 - n)\ dn.$$

$$\frac{dV}{dn} = 2(12 - n)$$

and for V to be a maximum $dV/dn = 0$.   Hence $2(12 - n) = 0$ and

$$n = 12$$

is the condition for a maximum value V.   That is, the meat should be
held twelve weeks in order that the net value shall be the greatest.

43. Problems for Solution.

1. Divide the number 48 into two parts such that the sum of the square
of one and the cube of the other shall be the least possible.

2. Divide the number 20 into two parts such that the sum of four times

the reciprocal of one and nine times the reciprocal of the other shall be the least possible.

3. A rectangular box with a square base and open at the top is to be constructed with a capacity of 500 cubic inches. What must be its dimensions to require the least material?

4. What are the most economical dimensions of an open cylindrical water tank if the cost of the curved side per square foot is two-thirds the cost of the bottom per square foot?

5. Find the altitude of the rectangle of greatest area that can be inscribed in a circle of radius R.

6. A vessel is anchored three miles off shore. Opposite a point five miles farther along the straight shore another vessel is anchored nine miles off. A boat from the first vessel is to land a passenger on the shore and then proceed to the second. What is the shortest total distance the boat can travel?

7. The speed of waves of length L in deep water is proportional to $\sqrt{\dfrac{L}{a} + \dfrac{a}{L}}$ where a is a certain constant. Show that the speed is the lowest when $L = a$.

8. From a certain circular sheet of metal a central sector is to be cut out so that the remainder, when the cut edges are drawn together, shall form a conical funnel top of the maximum capacity. Find the angle of the sector.

9. Find the height from the floor of a light on a wall so as to best illuminate a point on the floor b feet from the wall, assuming that the illumination varies directly as the sine of the angle of inclination of the rays to the floor and inversely as square of the distance from the light.

10. The strength of a beam of rectangular cross section varies directly as the breadth and the square of the depth. Find the shape of such a beam of the greatest strength which can be cut from a round log two feet in diameter.

11. The cost of fuel consumed in driving a certain ship through the water varies as the cube of the speed and is 25 dollars per hour when the speed is ten miles per hour. The other expenses are 100 dollars per hour, a fixed sum. What is the most economical speed?

12. A telephone company agrees to put in a new exchange for one hundred subscribers or less at a uniform charge of twenty dollars each. To encourage a large list of subscribers they agree to deduct ten cents from this uniform charge for each subscriber for every subscriber in excess of the hundred; for example, if there are 110 subscribers the deduction is one dollar and the charge nineteen dollars. For what number of subscribers would the company collect the greatest amount for the new exchange?

13. A farmer is located twelve miles from the nearest point on a straight railway. The railroad company agrees to put in a siding at any place he designates and haul his produce to a town eighty miles from his nearest point on the road for five cents a ton per mile. If he can haul by wagon to the railroad for thirteen cents a ton per mile on a straight road to the railroad, where should he have the siding placed in order that his transportation costs shall be the lowest per ton to the town?

14. The lower corner of the page of a book is folded over so as to just reach the inner edge of the page. Find the width of the part folded over, measured along the bottom of the page, when the length of the crease is the least possible.

15. A rectangular flower bed is to contain 432 sq. ft. It is surrounded by a walk which is r feet wide along the sides and 3 ft. wide across the ends. If the total area of the bed and walk together is a minimum, what are the dimensions of the bed?

16. A rectangular field to contain 3200 sq. rods is to be fenced off along the bank of a straight river. If no fence is needed along the river bank, what must be the dimensions to require the least amount of fencing?

17. A trough is to be made of a long rectangular piece of tin by bending up the two long edges so as to give a rectangular cross section. If the width of the piece is 25 inches, how deep should the trough be made in order that its carrying capacity may be a maximum?

18. A rectangular bin with a square base and cover is to be built to contain 1000 cu. ft. If the cost of the material for the bottom is 25 cents per sq. ft., for the top 40 cents, and for the sides 20 cents, what are the dimensions for minimum cost of material?

19. A man having 120 feet of fence wire wishes to enclose a rectangular yard and also build a fence across the yard parallel to two of the sides, dividing it into two smaller yards. What is the maximum area he can so fence off with the wire he has?

20. A window is in the form of a rectangle surmounted by an isosceles triangle, the altitude of the triangle being $\frac{3}{8}$ of its base. If the perimeter of the window is 30 feet, find the dimensions for admitting the greatest amount of light.

21. An electric current flows through a coil of radius r and exerts a force on a small magnet, the axis of which is on a line drawn through the center of the coil and perpendicular to its plane. This force is given by the formula $F = x/(r^2 + x^2)^{5/2}$, where x is the distance to the magnet from the center of the coil. Show that the force F is maximum when $x = \frac{1}{2}r$.

22. The turning effect of a ship's rudder is known to be $T = k \cos \theta \cdot \sin^2 \theta$, where k is a constant and θ is the angle which the direction of the rudder makes with the keel line. For what value of θ is the rudder most effective?

23. If a projectile is fired from a point O so as to strike an inclined plane which makes a constant angle A with the horizontal at O, the range is given by the formula $R = 2v^2 \cos \theta \cdot \sin (\theta - A)/g \cdot \cos^2 A$, where v and g are constants and θ is the angle of elevation of the gun. What value of θ gives the maximum range up the plane?

24. For a square-thread screw of pitch angle θ and friction angle ϕ the efficiency E is given by the formula $E = \tan \theta/[\tan (\theta + \phi) + f]$, where f is a constant. Find the value of θ for maximum efficiency when ϕ is a known constant angle.

25. A body of weight W is dragged along a horizontal plane by a force P whose line of action makes an angle θ with the plane. The magnitude of the force is given by the formula $P = mW/(m \sin \theta + \cos \theta)$, where m is the coefficient of friction. Show that when the pull is least $\tan \theta = m$.

Chapter 10

DIFFERENTIALS OF LOGARITHMIC AND EXPONENTIAL FUNCTIONS

44. Logarithmic and Exponential Functions. In article 9 the *logarithmic function* of a variable x is defined as $\log_b x$, where b is the constant base to which the logarithms are referred. Some of the theory of logarithms is studied in algebra, where it is found that although any number may be used as base, only two such are in general use. These are the so-called common or decimal logarithms with the base 10, which are used in ordinary numerical computations, and the so-called natural or hyperbolic logarithms which are used in theoretical formulations and calculations. In this chapter we find the differential of the logarithmic function with any base and see why one particular system is called the natural system.

For use in these derivations we recall from algebra three important properties of logarithms. To any permissible base the logarithm of a product of factors is the sum of the logarithms of the separate factors to the same base. Thus if N and n are any two positive real numbers and b represents any permissible base,

$$\log_b (Nn) = \log_b N + \log_b n. \tag{50}$$

Also, the logarithm of any power of any positive real number is equal to the logarithm of the number multiplied by the exponent of the power:

$$\log_b (N^n) = n(\log_b N). \tag{51}$$

Finally, if we know the logarithm of a number to a certain base and wish to find its logarithm referred to any other base, we simply multiply the known logarithm by the logarithm of the old base to the new base. Thus, if we know $\log_b N$ and wish to find $\log_a N$ the conversion formula is

$$\log_a N = (\log_a b) \log_b N. \tag{52}$$

From the definition of a logarithm as an exponent, as given in algebra, we know that if

100

$$y = \log_b x, \quad x = \mathfrak{z}^y. \tag{53}$$

The expression b^y in which b is constant and y variable was defined in article 9 as an *exponential function* of y. According to the definition of an inverse function in article 9 we see at once that the logarithmic and exponential functions are *inverse functions* of one another. Obviously in exponential functions the base as well as the exponent may be variable, as z^y, in which case the exponential is a function of two variables, y and z.

If in equations (53) the logarithm is the natural logarithm, the base is denoted by the letter e, and e^y is called *the* exponential function and is sometimes written *exp y*. If the variable is x we would write $e^x = $ exp x. This form of notation corresponds to that used for the trigonometric or circular functions, thus:

$$\sin x, \cos x, \exp x.$$

We proceed to find the differentials of the logarithmic and exponential functions, beginning with the logarithm.

45. Differential of the Logarithm of a Variable. Let x represent the variable and b the base of the logarithm. Then we have to find $d(\log_b x)$ and from this we at once get the rate and the derivative. Let z be another variable such that

$$z = mx; \quad \therefore \quad \log_b z = \log_b m + \log_b x,$$

where m is a constant and the second equation is obtained by applying equation (50). Differentiating these two expressions we have, since m and $\log_b m$ are constants,

$$dz = m\, dx, \quad d(\log_b z) = d(\log_b x).$$

Dividing the second of these two equations by the first,

$$\frac{d(\log_b z)}{dz} = \frac{1}{m} \cdot \frac{d(\log_b x)}{dx},$$

and if we multiply this equation by our first equation $z = mx$, member by member, the m cancels and we have

$$z \cdot \frac{d(\log_b z)}{dz} = x \cdot \frac{d(\log_b x)}{dx}. \tag{54}$$

The fractions on the two sides of this equation are the derivatives of $\log_b z$ and $\log_b x$, respectively, and since the constant m, which originally connected z and x, has been eliminated, z and x may be any two variables, related or unrelated. Equation (54), therefore, states that the product of any variable by the derivative of its logarithm is the same for any two variables, and hence the same for all variables. This product is then constant and we can write for any variable, say x,

$$x \cdot \frac{d(\log_b x)}{dx} = B,$$

where B is a constant whose value is at present unknown. From this,

$$d(\log_b x) = B \cdot \frac{dx}{x} \tag{55}$$

and in order to finally determine $d(\log_b x)$ we must find the value of B or express it in terms of the known constant base b.

Obviously the value of B will depend on that of the base b. If we use some other base a, we will have

$$d(\log_a x) = A \cdot \frac{dx}{x} \tag{56}$$

where A is another unknown constant whose value will depend on that of the known constant base a.

Divide equation (56) by (55) to eliminate the expression dx/x and obtain a relation between A, B and the differentials of the logarithms to the bases a, b. This gives

$$\frac{d(\log_a x)}{d(\log_b x)} = \frac{A}{B}. \tag{57}$$

Next refer back to equation (52) and differentiate it. Since $\log_a b$ is constant this gives

$$d(\log_a x) = (\log_a b) \cdot d(\log_b x),$$

$$\frac{d(\log_a x)}{d(\log_b x)} = \log_a b.$$

Comparing this equation with (57) we see that $\dfrac{A}{B} = \log_a b$. Hence

$$A = B(\log_a b),$$

and comparing this equation with equation (51) we see that

$$A = \log_a (b^B).$$

From this, by equations (53), $a^A = b^B$, and since both a and A are constants a^A is a constant. If we denote this constant by the letter e then $e = b^B$ and hence by equations (53)

$$B = \log_b e.$$

Substituting this value of B in formula (55) we have finally

$$d(\log_b x) = (\log_b e) \frac{dx}{x}. \qquad \text{(T)}$$

46. Natural Logarithms and Their Differentials.   In formula (T) the base b may be any number whatever.   *If we take* $b = e$ then $(\log_b e) = \log_e e$ and from algebra we know that the logarithm of any number referred to itself as base is 1.   If therefore we use the base e in formula (T) it becomes

$$d(\log_e x) = \frac{dx}{x}. \qquad \text{(U)}$$

The number e is therefore the base which gives the *simplest expression for the differential of the logarithm.*   For this reason e is called the *natural* base and logarithms to base e are called *natural logarithms.*   The number e is an unending decimal number and its value is 2.71828 + ⋯. We shall calculate this value to ten decimal places in a later chapter.

To convert logarithms from any base b into natural logarithms we use formula (52) with the base a replaced by e.   This gives

$$\log_e N = (\log_e b) \log_b N.$$

If the base b is 10 the formula for conversion from common to natural logarithms is therefore

$$\log_e N = (\log_e 10) \log_{10} N. \qquad \text{(58)}$$

The conversion factor $\log_e 10$ is also an unending decimal number and its value is 2.3026 + ⋯.   In a later chapter we shall also calculate this number to ten decimal places.

47. Differential of the Exponential Function of a Variable.   Let x be the variable and b the base, and let y be a dependent variable such that $y = b^x$ Then

$$\log_b y = x,$$

and we have to find $d(b^x)$ from these two relations.   Differentiating the second equation by formula (T) we have

$$(\log_b e) \frac{dy}{y} = dx.$$

$$\therefore \quad dy = \frac{1}{\log_b e} \cdot y \, dx.$$

But $y = b^x$, therefore this becomes

$$d(b^x) = \frac{1}{\log_b e} b^x\, dx. \tag{W}$$

Formula (W) gives the differential of an exponential function to any base. If we use the natural base e, then since $(\log_e e) = 1$, formula (W) becomes

$$d(e^x) = e^x\, dx. \tag{Y}$$

From this formula we have also, if a is any constant,

$$d(e^{ax}) = e^{ax} \cdot d(ax) = e^{ax} \cdot a\, dx,$$

$$\therefore \quad d(e^{ax}) = ae^{ax}\, dx. \tag{Z}$$

To find the rate of the exponential function we divide the differential by dt. Therefore by (Y),

$$\frac{d(e^x)}{dt} = e^x \frac{dx}{dt}. \tag{59}$$

The derivative is found by dividing (Y) by dx. Therefore

$$\frac{d(e^x)}{dx} = e^x. \tag{60}$$

Thus the exponential function (natural base) has the remarkable property that its derivative is equal to the function itself, and its rate is equal to the function itself multiplied by the rate of the exponent.

Since the first derivative is equal to the function, then the second derivative is equal to the first and therefore to the function. The same is true of the third and all its other derivatives. That is,

$$\frac{d^n}{dx^n}\, (e^x) = e^x. \tag{61}$$

These and other properties of the exponential and logarithmic functions will be investigated in a later chapter, together with their connection with other functions and other branches of mathematics. We shall also see that they have most important technical applications.

48. Illustrative Examples. 1. Find the differential of $\log_e (2x^3 + 3x^2)$.

Solution. By formula (U)

$$d[\log_e (2x^3 + 3x^2)] = \frac{d(2x^3 + 3x^2)}{2x^3 + 3x^2}. \tag{a}$$

But

$$d(2x^3 + 3x^2) = d(2x^3) + d(3x^2) = 6x^2\, dx + 6x\, dx$$
$$= 6x(x + 1)\, dx.$$

Also in the denominator of (a),

$$2x^3 + 3x^2 = x^2(2x + 3).$$

$$\therefore \quad d[\log_e (2x^3 + 3x^2)] = \frac{6x(x + 1)\, dx}{x^2(2x + 3)} = \frac{6(x + 1)}{x(2x + 3)}\, dx.$$

2. Differentiate $x \log_e x$.

Solution. This is the product of x and $\log_e x$.   Therefore

$$d(x \cdot \log_e x) = x \cdot d(\log_e x) + \log_e x \cdot d(x)$$

$$= x \cdot \frac{dx}{x} + \log_e x \cdot dx = dx + \log_e x \cdot dx$$

$$= (1 + \log_e x)\, dx.$$

3. Find $d[\log_{10} (3x + 2)]$.

Solution.   By formula (T), with $b = 10$,

$$d[\log_{10} (3x + 2)] = (\log_{10} e) \frac{d(3x + 2)}{3x + 2}.$$

But

$$d(3x + 2) = d(3x) = 3\, dx,$$

and

$$\log_{10} e = \log_{10} 2.7183 = .4343.$$

$$\therefore \quad d[\log_{10} (3x + 2)] = (.4343) \frac{3\, dx}{3x + 2} = \frac{1.3029}{3x + 2}\, dx.$$

4. If $y = \log_e \left(\dfrac{3x + 1}{x + 2} \right)$ find dy.

Solution.   By formula (U) for the natural logarithm,

$$dy = \frac{d\left(\dfrac{3x + 1}{x + 2} \right)}{\left(\dfrac{3x + 1}{x + 2} \right)} = \frac{x + 2}{3x + 1} \cdot d\left(\frac{3x + 1}{x + 2} \right).$$

But by the formula for the differential of a fraction,

$$d\left(\frac{3x + 1}{x + 2} \right) = \frac{(x + 2) \cdot d(3x + 1) - (3x + 1) \cdot d(x + 2)}{(x + 2)^2}$$

$$= \frac{(x + 2) \cdot 3\, dx - (3x + 1) \cdot dx}{(x + 2)^2}$$

$$= \frac{[3(x+2) - (3x+1)]\,dx}{(x+2)^2} = \frac{5}{(x+2)^2}\,dx.$$

$$\therefore \quad dy = \frac{x+2}{3x+1} \cdot \frac{5}{(x+2)^2}\,dx = \frac{5\,dx}{(x+2)(3x+1)}$$

$$= \frac{5\,dx}{3x^2 + 7x + 2}.$$

5. Differentiate $y = \frac{1}{2}e^{2x} + \frac{1}{3}e^{-3x}$.

Solution. $dy = d(\frac{1}{2}e^{2x}) + d(\frac{1}{3}e^{-3x}) = \frac{1}{2} \cdot d(e^{2x}) + \frac{1}{3} \cdot d(e^{-3x})$.
But by formula (Z),

$$d(e^{2x}) = 2e^{2x}\,dx,$$

$$d(e^{-3x}) = -3e^{-3x}\,dx.$$

$$\therefore \quad dy = \frac{1}{2} \cdot 2e^{2x}\,dx + \frac{1}{3}(-3e^{-3x}\,dx)$$

$$= (e^{2x} - e^{-3x})\,dx.$$

6. Find the derivative of $10^{.4343x}$.

Solution. By formula (W),

$$d(10^{.4343x}) = \frac{1}{\log_{10} e} \cdot 10^{.4343x} \cdot d(.4343x)$$

$$= \frac{1}{.4343} \cdot 10^{.4343x} \cdot .4343\,dx$$

$$= 10^{.4343x}\,dx.$$

$$\frac{d(10^{.4343x})}{dx} = 10^{.4343x}.$$

7. Find the greatest value of $\dfrac{\log_e (x^2)}{x^2}$.

Solution. Let y equal the given expression. Then for y to be a maximum $dy/dx = 0$.

Now in the numerator of the given expression, $\log_e (x^2) = 2 \log_e x$.

$$\therefore \quad y = 2\frac{\log_e x}{x^2}.$$

Differentiating,

$$dy = 2 \cdot d\left(\frac{\log_e x}{x^2}\right), \tag{a}$$

and since $\dfrac{\log_e x}{x^2}$ is a fraction

$$d\left(\frac{\log_e x}{x^2}\right) = \frac{x^2 \cdot d(\log_e x) - \log_e x \cdot d(x^2)}{(x^2)^2}$$

$$= \frac{x^2 \cdot \dfrac{dx}{x} - \log_e x \cdot 2x \, dx}{x^4}$$

$$= \frac{x(1 - 2\log_e x)\,dx}{x^4} = \frac{1 - 2\log_e x}{x^3}\,dx.$$

This result in (a) gives

$$dy = 2\left(\frac{1 - 2\log_e x}{x^3}\right)dx.$$

$$\therefore \quad \frac{dy}{dx} = 2\left(\frac{1 - 2\log_e x}{x^3}\right).$$

Hence for $dy/dx = 0$ we have the fraction on the right of this equation equal to zero, and for a fraction to be zero its numerator is zero. Therefore

$$1 - 2\log_e x = 0$$

$$2\log_e x = 1, \quad \text{or} \quad \log_e (x^2) = 1.$$

$$\therefore \quad x^2 = e$$

is the condition for y to be a maximum.   Substituting this value of x^2 in the original expression we get

$$y_{\text{max.}} = \frac{\log_e e}{e} = \frac{1}{e}.$$

8. In many branches of applied mathematics the functions

$$u = \tfrac{1}{2}(e^x + e^{-x}), \quad v = \tfrac{1}{2}(e^x - e^{-x})$$

are of great importance.   Show that $du/dx = v$ and $dv/dx = u$.

   Solution.

$$du = \tfrac{1}{2} \cdot d(e^x + e^{-x}) = \tfrac{1}{2}[d(e^x) + d(e^{-x})]$$

$$= \tfrac{1}{2}(e^x\,dx - e^{-x}\,dx) = \tfrac{1}{2}(e^x - e^{-x})\,dx.$$

$$\frac{du}{dx} = \tfrac{1}{2}(e^x - e^{-x}) = v.$$

Similarly,

$$dv = \tfrac{1}{2} \cdot d(e^x - e^{-x}) = \tfrac{1}{2}[d(e^x) - d(e^{-x})]$$

$$= \tfrac{1}{2}(e^x\,dx + e^{-x}\,dx) = \tfrac{1}{2}(e^x + e^{-x})\,dx$$

$$\frac{dv}{dx} = \tfrac{1}{2}(e^x + e^{-x}) = u.$$

10. In the disintegration of radium it is found by experiment that the amount present at any time after the beginning of the process can be expressed as a negative exponential function of the time. Show that the rate of disintegration at any instant is proportional to the amount of radium present at that time.

Solution. Let A be the amount remaining after t seconds. Then to express this as a negative exponential of the time we have

$$A = e^{-at}$$

where a is constant. The rate of disintegration (rate of change of A) is

$$\frac{dA}{dt} = -ae^{-at},$$

or

$$-\frac{dA}{dt} = ae^{-at},$$

the negative rate showing that A, the amount present, decreases.

But e^{-at} is A. Therefore the value of the negative rate is

$$\frac{dA}{dt} = aA,$$

which shows that it is proportional to A, the amount present, the constant of proportionality being a.

Exercises and Problems.

In each of the following find dy and $\dfrac{dy}{dx}$.

1. $y = \dfrac{1}{\log_e x}$.

2. $y = x^n \log_e x$.

3. $y = \log_e \left(\dfrac{ax - b}{ax + b} \right)$.

4. $y = b^x e^x$.

5. $y = (e^{3x} + 1)^3$.

6. $y = x^5 5^x$.

7. $y = \log_e (\log_e x)$.

8. $y = \log_e (x + \sqrt{x^2 - 1})$.

In each of the following find maximum and minimum values of y:

9. $y = \frac{1}{2}(e^x + e^{-x})$.

10. $y = \log_e (1 + x^2)$.

11. $y = x \log_e x$.

12. $y = \dfrac{x}{\log_e x}$.

13. What is the minimum value of $y = ae^{kx} + be^{-kx}$; a, b and k being constants?

14. A submarine cable consists of a core of copper wires with a covering

made of non-conducting material. If x denotes the ratio of the radius of the core to the thickness of the covering ($x = r/t$) it is proved in the theory of the submarine cable that the speed of signaling over the cable is given by the formula $S = kx^2 \log_e (1/x)$. Show that the maximum speed is attained when $x = 1/\sqrt{e}$.

15. The graph of the equation $y = e^{(-x^2)}$ is called the "probability curve," as it is important in the theory of probability and statistics. Find the maximum ordinate of this curve.

Chapter 11

SUMMARY OF
DIFFERENTIATION FORMULAS

49. Remarks. In the preceding chapters we have studied the mathematical meaning of rates of change of varying quantities and found that this could be expressed by means of formulas and equations giving the differentials of the quantities. We then derived formulas for all the ordinary forms of expressions, or functions.

From these differential formulas the rates or derivatives are immediately obtained and we have used these in working many problems which involved the rates of change of certain quantities under specified conditions and also the greatest and least values they could have under specified conditions. We shall have need for all of these formulas in our later work in which we examine the question of varying quantities from another viewpoint.

For reference and review, therefore, we collect in this chapter a list of all the differential formulas and the corresponding derivatives. These are given in the next article.

50. Differential and Derivative Formulas.

	Function	Differential	Derivative
(A)	$x + y + z + \cdots$	$dx + dy + dz + \cdots$	
(B)	$y = c$	$dy = 0$	$dy/dx = 0$
(C)	$y = x + c$	$dy = dx$	$dy/dx = 1$
(D)	$y = \pm x$	$dy = \pm dx$	$dy/dx = \pm 1$
(E)	$y = mx$	$dy = m\, dx$	$dy/dx = m$
(F)	$y = x^2$	$dy = 2x\, dx$	$dy/dx = 2x$
(G)	$y = \sqrt{x}$	$dy = \dfrac{dx}{2\sqrt{x}}$	$dy/dx = \dfrac{1}{2\sqrt{x}}$
(H)	$z = xy$	$dz = x\, dy + v\, dx$	
(J)	$y = \dfrac{1}{x}$	$dy = -\dfrac{dx}{x^2}$	$dy/dx = -\dfrac{1}{x^2}$
(K)	$z = \dfrac{x}{y}$	$dz = \dfrac{y\, dx - x\, dy}{y^2}$	

110

Function	Differential	Derivative
(L) $y = x^n$	$dy = nx^{n-1}\,dx$	$dy/dx = nx^{n-1}$
(M) $y = \sin\theta$	$dy = \cos\theta\,d\theta$	$dy/d\theta = \cos\theta$
(N) $y = \cos\theta$	$dy = -\sin\theta\,d\theta$	$dy/d\theta = -\sin\theta$
(P) $y = \tan\theta$	$dy = \sec^2\theta\,d\theta$	$dy/d\theta = \sec^2\theta$
(Q) $y = \cot\theta$	$dy = -\csc^2\theta\,d\theta$	$dy/d\theta = -\csc^2\theta$
(R) $y = \sec\theta$	$dy = \sec\theta\tan\theta\,d\theta$	$dy/d\theta = \sec\theta\tan\theta$
(S) $y = \csc\theta$	$dy = -\csc\theta\cot\theta\,d\theta$	$dy/d\theta = -\csc\theta\cot\theta$
(T) $y = \log_b x$	$dy = (\log_b e)\dfrac{dx}{x}$	$dy/dx = \dfrac{1}{x}(\log_b e)$
(U) $y = \log_e x$	$dy = \dfrac{dx}{x}$	$dy/dx = \dfrac{1}{x}$
(W) $y = b^x$	$dy = \dfrac{b^x\,dx}{\log_b e}$	$dy/dx = \dfrac{b^x}{\log_b e}$
(Y) $y = e^x$	$dy = e^x\,dx$	$dy/dx = e^x$
(Z) $y = e^{ax}$	$dy = ae^{ax}\,dx$	$dy/dx = ae^{ax}$

This list is sometimes called the *differential catechism* and is usually contained in the handbooks of engineering, physics and chemistry. One who expects to use the calculus frequently should master this list thoroughly. The following list of exercises will provide practice in the use of the formulas.

Review Exercises.

Find the differential and the derivative of each of the following functions:

1. $y = x^3 - 3x$.

2. $y = \dfrac{x}{a} + \dfrac{a}{x}$.

3. $y = \sqrt{ax + b}$.

4. $y = x\sqrt{a^2 - x^2}$.

5. $s = ae^{bt}$.

6. $u = \log_e (cv)$.

7. $r = \sin (a\theta)$.

8. $y = \log_e (\sin x)$.

9. $r = \theta \cdot \cos\theta$.

10. $s = e^t \cos t$.

11. $y = \sqrt{\dfrac{x}{a}} - \sqrt{\dfrac{a}{x}}$.

12. $u = \sqrt{e^v + 1}$.

13. $y = \dfrac{x}{\sqrt{a^2 - x^2}}$.

14. $y = \sqrt{\dfrac{a - x}{a + x}}$.

15. $r = 2\sin\left(\dfrac{\theta}{2}\right)$.

16. $s = e^{-at}\sin bt$.

17. $r = \sqrt{\cot\theta}$.

18. $y = \log_e \sqrt[3]{\dfrac{6x - 5}{4 - 3x}}$.

19. If $x^2 + y^2 = a^2$, show that $dy = -\dfrac{x}{y}\,dx$.

20. If $\sqrt{x} + \sqrt{y} = \sqrt{a}$, show that $dy = -\sqrt{\dfrac{y}{x}}\,dx$.

Chapter 12

REVERSING THE PROCESS
OF DIFFERENTIATION

51. A New Type of Problem. In all our work so far we have had given, or could formulate directly from given data, the relation between two quantities and were required to find the rate of one in terms of the rate of the other, or to find the greatest or least value of one corresponding to certain values of the other. Either of these problems depended on our finding the rate or derivative of one of the quantities in terms of the other, which in turn required that the differentials be found.

Let us suppose now that the problem is reversed. This means that we have given, or can immediately formulate from given data, the rate or derivative of one quantity in terms of the other and are required to find the relation between the quantities themselves, and this in turn will depend on finding the quantity when we know its differential.

As an illustration of the first problem let us recall the example of the falling body which we discussed in articles 29 and 30. Here it is known by experiment that in a specified time t seconds an object falls through a space s feet which is related to the time by the equation

$$s = 16t^2 \tag{62}$$

and we are required to find (i) the velocity which it has at the end of the time t, or (ii) the acceleration which it must have in order to attain that velocity. Since the velocity v equals ds/dt, the first requires us to find the rate of s in terms of t from the relation (62), that is, the derivative of s with respect to t, ds/dt. Since the acceleration a equals $\dfrac{d^2s}{dt^2}$, the second requires us to find the second derivative of s with respect to t, from the same relation (62).

The answer to the first is, as we know, $ds = 32t\,dt$ and

$$v = \frac{ds}{dt} = 32t, \tag{63}$$

and the answer to the second is $dv = d\left(\dfrac{ds}{dt}\right) = 32\ dt$ or

$$\frac{dv}{dt} = \frac{d^2s}{dt^2} = \frac{d}{dt}\left(\frac{ds}{dt}\right) = 32.$$

That is, at any time t the falling object has a velocity $v = 32t$, and to attain this velocity in the time t when falling freely its acceleration is $a = 32$.

If the problem is now reversed, we have given $a = \dfrac{dv}{dt} = 32$, (i) to find the velocity at any time t, and from the velocity $v = \dfrac{ds}{dt} = 32t$, (ii) to find the relation between s and t. The answer to the first is clear from the following: Since $dv/dt = 32$, then

$$dv = 32\ dt, \tag{64}$$

and we have to *find v when we know its differential*. From (64) we know, by reversing the process of differentiation, that

$$v = 32t, \tag{65}$$

which is the relation between v and t given by equation (63).

The answer to the second question is now also clear: Since $v = ds/dt$, we have $ds/dt = 32t = 16\cdot2t$, then

$$ds = 16(2t\ dt), \tag{66}$$

and again we have to *find s when we know its differential*. From (66) we know that, by again reversing the process of differentiation, $s = 16t^2$, which is the relation between s and t given by the original equation (62).

In the present case we are able to check our results in reversing the process of differentiation because we already have the original relations to which we were to work back. Suppose, however, we have a case in which we are given the derivative, rate or differential and do *not* know beforehand the required original relation between the quantities involved. How then shall we proceed? The answer is, of course, plain: Simply recognize the given expression as the result obtained by one of our differentiation formulas of article 50 and *reverse the process of differentiation*.

Thus, suppose we have given the fact that in a certain problem the derivative of a certain function is

$$\frac{dy}{dx} = 4x^3 \tag{67}$$

and are asked to find the original relation between y and x. By multiplying (67) by dx we have

$$dy = 4x^3\, dx,$$

and therefore, as we know from finding the differential of x^4, it must be that $dy = d(x^4)$, that is,

$$y = x^4,$$

and all we had to do was to *find y when its differential is known.*

52. Integration and Integrals. The considerations of the preceding article form the foundations of a second branch of the calculus. In order to formulate the principle brought out in those considerations let us summarize in convenient form what we have done.

In the first case with $a = dv/dt = 32$, or with

$$dv = 32\, dt$$

known we found

$$v = 32t.$$

That is, we reversed the differential formula (E) of the list in article 50, the number 32 corresponding to the constant coefficient m.

In the second case with $v = ds/dt = 32t$, or with

$$ds = 32t\, dt = 16(2t\, dt)$$

known, we found

$$s = 16t^2.$$

In this case we reversed the differential formula (F), there being also a constant coefficient, namely 16, prefixed to this formula.

In the case of $dy/dx = 4x^3$, or

$$dy = 4x^3\, dx,$$

we similarly reversed the formula (L) with $n = 4$ and $n - 1 = 3$.

It should now be plain that when we know the derivative or differential relation between two quantities, in order to find the relation between the quantities themselves, it is only necessary to recognize the differential as corresponding to one of the formulas of the differential catechism and then write out the corresponding function by reversing the differential formula. This process, the reverse of differentiation, is called *integration.* Also, as we find the differential when we differentiate, so in the process of integration we are said to *integrate* an expression or to find its *integral*.

Thus in the examples discussed above we

integrate 32 *dt*,   or   find its integral 32*t*;
integrate 16(2*t dt*),   find its integral 16*t*²;
integrate 4*x*³ *dx*,   find its integral *x*⁴.

Just as we have a symbol for differentiation and write

$$d(t) = dt$$

$$d(t^2) = 2t\, dt$$

$$d(x^4) = 4x^3\, dx,$$

so also we have a symbol for integration, and we write

$$\left.\begin{aligned} \int dt &= t \\[2mm] \int 2t\, dt &= t^2 \\[2mm] \int 4x^3\, dx &= x^4 \end{aligned}\right\} \tag{68}$$

The word *differential* comes from the "difference" or part of a quantity and the letter *d* is the initial.   Similarly the word *integral* means whole or total and the symbol $\int$ is an elongated *S*, the initial letter of the word "sum."   To differentiate a quantity or function is to find its differential or part, which results from a small change in value of the independent variable, and to integrate is to sum up the differentials, that is, find the original function or quantity from its differential. The part of the calculus which has to do with integrals and integration and their applications is called the *Integral Calculus*.

We shall later find a fuller interpretation of this process of integration and a fuller appreciation of its significance by applying it to geometrical and technical problems.   For the present we must concern ourselves with the matter of obtaining integral formulas such as those of (68) for all the ordinary functions, in the same way that we first had to find the differential formulas before we could handle problems in which rates and derivatives played a part.

53. Integral Formulas. In solving problems which involve the rates of change of quantities, the relation between given data and required result may take any form whatever, and therefore the formula or equation giving the rate may be any form of expression whatever. However, the relations between the quantities involved are nothing more than what we call functions, and we saw in article 9 that all functions can be divided into definite classes. Since the finding of rates and derivatives depends on knowing the differentials, it was only necessary to find the differentials of certain classes of functions, combinations of which go to make up almost all the functional relations usually met with.

We found that there are about two dozen such functions and their differentials are called the *standard forms*. When we know these standard forms we are able to differentiate any expression which may be met.

Similarly, if we can integrate a certain number of expressions which are met most frequently, we can, by combining these forms or by reducing a given expression to combinations of these forms, integrate the more usual expressions.

Since the operation of integration is the inverse of differentiation, we can obviously find a number of integrals simply by reversing the standard differential formulas. Since, however, some of the differential formulas were obtained by an indirect method, their reversal will require an indirect procedure. The list of integrals obtained in either way from the differential formulas and related formulas are also called *standard forms*, and we shall find also that there are about two dozen of them.

In the next chapter we take up the derivation of the standard integral formulas.

Chapter 13

INTEGRAL FORMULAS

54. Integral Formulas Obtained Directly from Differentials.
The references to differential formulas made here will be to the list in article 50 of Chapter 11.

Referring to formula (A) it is at once obvious that since

$$d(x + y + z + \cdots) = dx + dy + dz + \cdots$$

then the reverse operation must also deal with the entire expression one term at a time. That is,

$$\int (dx + dy + dz + \cdots) = \int dx + \int dy + \int dz + \cdots \qquad \text{(I)}$$

and so for any expression made up of the sum or differences of terms of any form, to integrate the entire expression integrate each term separately.

In formula (C) the process of differentiation caused the constant c to disappear from the result, because by formula (B) a quantity which does not vary has no differential. In the reverse process, therefore, the constant c must reappear. That is, from (C)

$$\int dx = x + C. \qquad \text{(II)}$$

This simple-looking formula is of the greatest importance; it means that in any integration a constant must be added to the result. The reason for this is, of course, that since we do not know before we integrate whether there was an added constant in the original function or not, we should always include it to be certain. In applied problems the conditions of the problem will determine whether or not the added constant is to be retained in any case, and if so, determine its value.

Later on we shall discuss this matter further, but for the present it is to be understood that in every case a constant is to be added after integrating. This will apply to formula (I) above and to every formula which follows, even though for simplicity in writing we may not always include it.

Reversing formula (D) and adding the constant of integration we get at once

$$\int (\pm dx) = \pm \int dx = \pm x + C. \tag{III}$$

This formula states that the integral of a negative expression is negative, and that of a positive expression is positive.

In formula (E), the process of differentiation does not apply to the constant m, since a constant has no differential, but only to the variable x. Therefore, in integration the constant m may be removed "outside" the integral sign, that is, placed before it, and we write

$$\int m \, dx = m \int dx. \tag{IV}$$

$$\therefore \quad \int m \, dx = mx + C$$

will apply in every case in which we have to integrate an expression which has a constant coefficient. The constant coefficient can be removed outside the integral sign and only used again to multiply the integrated expression after the process is complete. It does not matter whether the variable part of the expression be simply dx or the differential of any single variable, or a complicated expression formed from the differential of another complicated function. The constant multiplier in either case does not enter into the integration, but remains as a simple multiplier. From these considerations it will be seen at once that we may introduce any multiplying or dividing constant inside the integral sign at will if we will at the same time do the reverse and divide or multiply by the same constant on the outside of the integral so as to cancel the effect inside.

In deriving the differential formulas we found that formulas (F), (G), and (J) are all equivalent to (L), because in (F), x^2 is x^n with $n = 2$, in (G) $\sqrt{x} = x^{1/2}$ or $n = \frac{1}{2}$, and in (J) $1/x = x^{-1}$ or $n = -1$. The integral formula corresponding to (L) will therefore correspond also to (F), (G) and (J).

The integral of (L) is at once seen to be

$$\int n x^{n-1} \, dx = x^n$$

but by (IV), we can write

$$\int nx^{n-1}\, dx = n \int x^{n-1}\, dx$$

since n is a constant. This would mean that we are to integrate $x^{n-1}\, dx$ without reference to the coefficient n. How are we to find $\int x^{n-1}\, dx$ alone without the coefficient when we know that the coefficient n is an essential part of the result? In other words, since if n is any real number whatever so is $n - 1$, how would we in general integrate a variable raised to any power? How would we integrate $x^3\, dx$ without the coefficient 4, which tells us that x^3 results from differentiating x^4? Using the letter m to represent any number whatever, how shall we integrate $x^m\, dx$?

To determine this let us find the differential of the variable raised to a power 1 greater than m, that is, find $d(x^{m+1})$. By the differential formula (L)

$$d(x^{m+1}) = (m + 1)\cdot x^{(m+1)-1}\, dx$$
$$= (m + 1)x^m\, dx.$$

Dividing both sides of this equation by $(m + 1)$, and transposing we have,

$$x^m\, dx = \left(\frac{1}{m + 1}\right)\cdot d(x^{m+1}),$$

$$\therefore \quad \int x^m\, dx = \int \left(\frac{1}{m + 1}\right)\cdot d(x^{m+1}).$$

However, since m is a constant, so is $1/(m + 1)$. Therefore, we can place $1/(m + 1)$ outside the integral sign according to formula (IV), and write the last equation as

$$\int x^m\, dx = \left(\frac{1}{m + 1}\right)\cdot \int d(x^{m+1}).$$

But according to formula (II) $\int d(x^{m+1}) = x^{m+1} + C$. Therefore

$$\int x^m\, dx = \left(\frac{1}{m + 1}\right)\cdot x^{m+1} + C,$$

or, for simplicity in writing,

$$\int x^m\, dx = \frac{x^{m+1}}{m + 1}. \tag{V}$$

This is the final formula for integrating $x^m\, dx$ in which m may represent any exponent whatever (except $m = -1$) as n may represent any exponent whatever in the differentiation formula (L).

Let us see now whether or not formula (V) is the reverse of formula (L). In (L), to differentiate x to any power we *decrease* the exponent by 1 and *multiply* by the *old* exponent. In (V), to integrate x to any power we *increase* the exponent by 1 and *divide* by the *new* exponent. Formula (V) is therefore exactly the reverse of formula (L).

Let us see now if formula (V) will give us the reverse of differentiation formulas (F), (G) and (J). If so, then we should get

$$\int 2x\, dx = x^2, \quad \int \frac{dx}{2\sqrt{x}} = \sqrt{x}, \quad \int -\frac{dx}{x^2} = \frac{1}{x}.$$

Since 2 is constant and x is x^1, we can write the first of these three as $2\int x^1\, dx$. Then, according to formula (V), we get

$$2\left[\int x^1\, dx\right] = 2\left[\frac{x^{1+1}}{1+1}\right] = 2\left(\frac{x^2}{2}\right) = x^2.$$

Also, $\dfrac{1}{2\sqrt{x}} = \dfrac{1}{2}\dfrac{1}{x^{1/2}} = \dfrac{1}{2} x^{-1/2}$. Therefore,

$$\int \frac{dx}{2\sqrt{x}} = \int \frac{1}{2} x^{-1/2}\, dx = \frac{1}{2}\int x^{-1/2}\, dx = \frac{1}{2}\left(\frac{x^{-1/2+1}}{-\frac{1}{2}+1}\right)$$

and

$$\frac{1}{2}\left(\frac{x^{-1/2+1}}{-\frac{1}{2}+1}\right) = \frac{1}{2}\left(\frac{x^{1/2}}{\frac{1}{2}}\right) = x^{1/2} = \sqrt{x}.$$

For the third, since the minus sign goes outside the integral sign by formula (III) and $\dfrac{dx}{x^2} = x^{-2}\, dx$, then by (V)

$$\int -\frac{dx}{x^2} = -\int x^{-2}\, dx = -\left(\frac{x^{-2+1}}{-2+1}\right) = -\left(\frac{x^{-1}}{-1}\right) = x^{-1} = \frac{1}{x}.$$

Therefore, integration formula (V) is precisely the reverse of (L) and includes the reverse process of each of (F), (G) and (J).

Formula (M) shows at once that

$$\int \cos \theta\, d\theta = \sin \theta \qquad\qquad \text{(VII)}$$

but by (N) $d(\cos \theta) = -\sin \theta \, d\theta$, hence on reversing this formula,

$$\int d(\cos \theta) = \int (-\sin \theta \, d\theta).$$

$$\therefore \quad \cos \theta = -\int \sin \theta \, d\theta,$$

or,

$$\int \sin \theta \, d\theta = -\cos \theta. \tag{VI}$$

In the same way we get from the differential formulas (P), (Q), (R) and (S) the integral formulas

$$\int \sec^2 \theta \, d\theta = \tan \theta \tag{XII}$$

$$\int \csc^2 \theta \, d\theta = -\cot \theta \tag{XIII}$$

$$\int \sec \theta \tan \theta \, d\theta = \sec \theta \tag{XIV}$$

and

$$\int \csc \theta \cot \theta \, d\theta = -\csc \theta. \tag{XV}$$

By reversing differential formula (U) we get at once

$$\int \frac{dx}{x} = \log_e x. \tag{XVI}$$

This is an important formula and will often be referred to.

Digressing a moment, we may say, in connection with formula (XVI), that integration is a process which can be defined without any reference to or knowledge of logarithms, and when integration is once defined, the above integral can be taken as the definition of a logarithm. This is sometimes done and the definition takes the following form: "The logarithm of a number z to the base e is a number L such that

$$L = \int \frac{dz}{z}. \tag{69}$$

This is written, $L = \log_e z$." After this definition is given, it is then easy to show from the differential formulas involving $\log_e z$ that $z = e^L$ and from this that for any base b if $l = \log_b z$ then $z = b^l$. These definitions, therefore, amount to the same as the ordinary definition of logarithms given in algebra.

Returning to our differential formulas, we also see at once that from formula (Y) we get directly

$$\int e^x \, dx = e^x. \tag{XVII}$$

In formula (Z) to get $d(e^{ax})$ we write e with the same exponent and *multiply* by the constant a. In reversing this to get the integral, we write e with the same exponent and *divide* by a. This gives us

$$\int e^{ax} \, dx = \frac{e^{ax}}{a}. \tag{XVIII}$$

This completes the list of integral formulas which are easily and directly obtained from the differential formulas. In the next article we derive several which require some simple transformations.

55. Integral Formulas Derived Indirectly. In our list of differentials there is no formula which gives $\tan \theta \, d\theta$ as the differential of another function. We cannot, therefore, find $\int \tan \theta \, d\theta$ directly by reversal of a differential formula. The same is true of $\cot \theta$, $\sec \theta$ and $\csc \theta$, and also of a few logarithmic and exponential functions. For this reason we must find these by certain simple transformations of the formulas which we do have.

Since $\tan \theta = \sin \theta / \cos \theta = -(-\sin \theta / \cos \theta)$, therefore

$$\int \tan \theta \, d\theta = -\int \left(\frac{-\sin \theta}{\cos \theta}\right) d\theta = -\int \frac{(-\sin \theta \, d\theta)}{\cos \theta}.$$

But $(-\sin \theta \, d\theta) = d(\cos \theta)$ by formula (N). Hence

$$\int \tan \theta \, d\theta = -\int \frac{d(\cos \theta)}{\cos \theta}.$$

But this integral is of the same form as formula (XVI) with x replaced by $\cos \theta$. Therefore

$$\int \tan \theta \, d\theta = -\log_e (\cos \theta) \tag{VIII}$$

Similarly

$$\int \cot \theta = \int \left(\frac{\cos \theta}{\sin \theta}\right) d\theta = \int \frac{(\cos \theta \, d\theta)}{\sin \theta} = \int \frac{d(\sin \theta)}{\sin \theta}.$$

$$\therefore \quad \int \cot \theta \, d\theta = \log_e (\sin \theta). \qquad \text{(IX)}$$

To find $\int \sec \theta \, d\theta$ we multiply $\sec \theta$ by $\dfrac{\tan \theta + \sec \theta}{\tan \theta + \sec \theta}$, which is equal to 1, and, of course, makes no change in the original expression except in its form. We get then

$$\int \sec \theta \, d\theta = \int \frac{\sec \theta (\tan \theta + \sec \theta) \, d\theta}{\tan \theta + \sec \theta},$$

or, by multiplying out the parentheses,

$$\int \sec \theta \, d\theta = \int \frac{(\sec \theta \tan \theta \, d\theta) + (\sec^2 \theta \, d\theta)}{\sec \theta + \tan \theta}.$$

But

$$d(\sec \theta + \tan \theta) = d(\sec \theta) + d(\tan \theta) = (\sec \theta \tan \theta \, d\theta) + (\sec^2 \theta \, d\theta).$$

Therefore, the numerator of the fraction is the differential of the denominator and we can write it

$$\int \sec \theta \, d\theta = \int \frac{d(\sec \theta + \tan \theta)}{\sec \theta + \tan \theta},$$

and this is also of the form of (XVI). Therefore

$$\int \sec \theta \, d\theta = \log_e (\sec \theta + \tan \theta). \qquad \text{(X)}$$

To find $\int \csc \theta \, d\theta$ we use the same kind of scheme and write it

$$\int \csc \theta \, d\theta = \int \frac{\csc \theta (-\cot \theta + \csc \theta) \, d\theta}{-\cot \theta + \csc \theta}$$

$$= \int \frac{(-\csc \theta \cot \theta \, d\theta) + \csc^2 \theta \, d\theta)}{\csc \theta - \cot \theta}$$

$$= \int \frac{d(\csc \theta - \cot \theta)}{\csc \theta - \cot \theta}.$$

$$\therefore \quad \int \csc \theta \, d\theta = \log_e (\csc \theta - \cot \theta). \qquad \text{(XI)}$$

We were able to find $\int e^x \, dx$ without any trouble because e^x is its own derivative. For any other exponential such as b^x, however, this is not the case. In order to find $\int b^x \, dx$ we must find a suitable transformation.

From formula (W) we have $d(b^x) = \left(\dfrac{1}{\log_b e}\right) b^x \, dx$, therefore,

$$b^x \, dx = (\log_b e) \cdot d(b^x)$$

and

$$\int b^x \, dx = \int (\log_b e) \cdot d(b^x) = (\log_b e) \int d(b^x),$$

since $\log_b e$ is a constant.

$$\therefore \quad \int b^x \, dx = (\log_b e) b^x. \tag{XIX}$$

From formula (H), if we replace x and y by u and v, we get

$$d(uv) = u \, dv + v \, du$$

and by integrating,

$$\int d(uv) = \int u \, dv + \int v \, du,$$

$$\therefore \quad uv = \int u \, dv + \int v \, du.$$

We cannot perform the two separate integrations on the right of this equation because in the first dv is not expressed in terms of u and in the second du is not expressed in terms of v. Transposing the equation, however,

$$\int u \, dv = uv - \int v \, du, \tag{70}$$

which is an important formula for use in transforming other integrals. Although it does not give us exactly the value of the integral $\int u \, dv$ on the left, the equation gives this integral partially by stating that it is equal to uv minus another integral. This process is sometimes called *partial integration* or integration by parts. If both the factors u and v of the product uv are known separately then du can be found and when this value is put in the second integral on the right the process can frequently be completed.

We shall use equation (70) to find $\int \log_e x \, dx$, which is not the direct result of differentiation of any single term and so cannot be obtained by the direct reversal of any of our differential formulas.

By formula (U) $d(\log_e x) = dx/x$, hence $x \cdot d(\log_e x) = dx$. Integrating this,

$$\int x \cdot d(\log_e x) = \int dx = x. \tag{71}$$

Now the integral $\int x \cdot d(\log_e x)$ corresponds exactly to the integral $\int u \cdot dv$ with x in place of u and $\log_e x$ in place of v. We can write, therefore,

$$u = x, \quad v = \log_e x, \quad \text{and} \quad du = dx.$$

If we put these values of u, v and du in equation (70) we have

$$\int x \cdot d(\log_e x) = x \cdot \log_e x - \int \log_e x \cdot dx.$$

But in this, by equation (71), $\int x \cdot d(\log_e x) = x$. Therefore,

$$x = x \cdot \log_e x - \int \log_e x \cdot dx.$$

Transposing this we have, $\int \log_e x \cdot dx = x \cdot \log_e x - x$, or, taking out x as a factor of the two terms on the right,

$$\int \log_e x \, dx = x(\log_e x - 1). \tag{XX}$$

In order to find the integral, $\int \log_b x \, dx$, we use the same method but start with differential formula (T). This formula states that $d(\log_b x) = (\log_b e) \cdot dx/x$. Therefore,

$$x \cdot d(\log_b x) = (\log_b e) \, dx.$$

$$\therefore \quad \int x \cdot d(\log_b x) = \int (\log_b e) \, dx$$
$$= (\log_b e) \cdot \int dx = (\log_b e) \cdot x, \tag{72}$$

($\log_b e$) being a constant. If now we let $u = x$, $v = \log_b x$, $du = dx$, in equation (70) we get for the first integral on the left of (72),

$$\int x \cdot d(\log_b x) = x \cdot \log_b x - \int \log_b x \cdot dx.$$

But by (72), the integral on the left here is equal to ($\log_b e) \cdot x$. Therefore,

$$(\log_b e) \cdot x = x \cdot \log_b x - \int \log_b x \cdot dx,$$

or

$$\int \log_b x \cdot dx = x \cdot \log_b x - x \cdot \log_b e = x(\log_b x - \log_b e).$$

But the difference between two logarithms is the logarithm of a quotient. Applying this principle to the expression in parentheses we get finally

$$\int \log_b x \, dx = x \log_b \left(\frac{x}{e}\right). \tag{XXI}$$

This completes the list of the more useful and usual integral formulas, though there are many more which might be added to the list.

56. Summary of Integrals. We collect here for easy reference the integrals found in this chapter. A longer list is usually given in the handbooks of engineering, physics and chemistry.

In using the following list it is to be understood that a constant is added after each result as in formula (II).

(I) $\quad \int (dx + dy + dz + \cdots) = \int dx + \int dy + \int dz \cdots$

(II) $\quad \int dx = x + C$

(III) $\quad \int (\pm dx) = \pm \int dx$

(IV) $\quad \int m \, dx = m \int dx$

(V) $\quad \int x^m \, dx = \dfrac{x^{m+1}}{m + 1} \quad (m \neq -1)$

(VI) $\quad \int \sin \theta \, d\theta = -\cos \theta$

(VII) $\displaystyle\int \cos \theta \, d\theta = \sin \theta$

(VIII) $\displaystyle\int \tan \theta \, d\theta = -\log_e (\cos \theta)$

(IX) $\displaystyle\int \cot \theta \, d\theta = \log_e (\sin \theta)$

(X) $\displaystyle\int \sec \theta \, d\theta = \log_e (\sec \theta + \tan \theta)$

(XI) $\displaystyle\int \csc \theta \, d\theta = \log_e (\csc \theta - \cot \theta)$

(XII) $\displaystyle\int \sec^2 \theta \, d\theta = \tan \theta$

(XIII) $\displaystyle\int \csc^2 \theta \, d\theta = -\cot \theta$

(XIV) $\displaystyle\int \sec \theta \tan \theta \, d\theta = \sec \theta$

(XV) $\displaystyle\int \csc \theta \cot \theta \, d\theta = -\csc \theta$

(XVI) $\displaystyle\int \frac{dx}{x} = \log_e x$

(XVII) $\displaystyle\int e^x \, dx = e^x$

(XVIII) $\displaystyle\int e^{ax} \, dx = \frac{e^{ax}}{a}$

(XIX) $\displaystyle\int b^x \, dx = (\log_b e) b^x$

(XX) $\displaystyle\int \log_e x \, dx = x(\log_e x - 1)$

(XXI) $\displaystyle\int \log_b x \, dx = x \log_b \left(\frac{x}{e}\right)$

57. Illustrative Examples.   1. If the differential of y is

$$dy = (4x^3 + 3x^2 + 2x + 1)\, dx$$

find the value of y.

Solution. The value of y is $\int dy$. Therefore

$$y = \int (4x^3 + 3x^2 + 2x + 1)\, dx,$$

or by formula (I),

$$y = \int 4x^3\, dx + \int 3x^2\, dx + \int 2x\, dx + \int dx,$$

and by (IV),

$$y = 4 \int x^3\, dx + 3 \int x^2\, dx + 2 \int x\, dx + \int dx.$$

Hence by (V),

$$y = 4 \left(\frac{x^4}{4}\right) + 3 \left(\frac{x^3}{3}\right) + 2 \left(\frac{x^2}{2}\right) + x + C.$$

$$\therefore \quad y = x^4 + x^3 + x^2 + x + C.$$

2. Integrate $2(\cos 2x - \sin 2x)$.

Solution.

$$\int [2(\cos 2x - \sin 2x)]\, dx = 2 \int \cos 2x\, dx - 2 \int \sin 2x\, dx.$$

Let $2x = \theta$, then $2\, dx = d\theta$ and $dx = \tfrac{1}{2}\, d\theta$. Therefore

$$\cos 2x\, dx = \cos \theta \cdot \tfrac{1}{2}\, d\theta = \tfrac{1}{2} \cos \theta\, d\theta,$$

and similarly for $\sin 2x\, dx$. Hence

$$\int \cos 2x\, dx = \int \tfrac{1}{2} \cos \theta\, d\theta = \tfrac{1}{2} \int \cos \theta\, d\theta = \tfrac{1}{2} \sin \theta,$$

by formula (VII);

$$\int \sin 2x\, dx = \int \tfrac{1}{2} \sin \theta\, d\theta = \tfrac{1}{2} \int \sin \theta\, d\theta = -\tfrac{1}{2} \cos \theta,$$

by formula (VI).

$$\therefore \; 2 \int \cos 2x \, dx - 2 \int \sin 2x \, dx = 2(\tfrac{1}{2} \sin \theta) - 2(-\tfrac{1}{2} \cos \theta)$$

$$= \sin \theta + \cos \theta = \sin 2x + \cos 2x.$$

$$\int [2(\cos 2x - \sin 2x)] \, dx = \sin 2x + \cos 2x + C.$$

3. Find the value of $\displaystyle \int \left(x^{2/3} - \frac{1}{x^{2/3}} + \frac{2}{x^3} - \frac{1}{x^2} \right) dx$.

Solution.　$x^{2/3} - \dfrac{1}{x^{2/3}} + \dfrac{2}{x^3} - \dfrac{1}{x^2} = x^{2/3} - x^{-2/3} + 2x^{-3} - x^{-2}.$

Therefore, the integral of the expression is

$$\int x^{2/3} \, dx - \int x^{-2/3} \, dx + 2 \int x^{-3} \, dx - \int x^{-2} \, dx$$

$$= \frac{x^{2/3+1}}{(\tfrac{2}{3} + 1)} - \frac{x^{-2/3+1}}{(-\tfrac{2}{3} + 1)} + 2 \frac{x^{-3+1}}{(-3 + 1)} - \frac{x^{-2+1}}{(-2 + 1)} + C$$

$$= \frac{x^{5/3}}{(\tfrac{5}{3})} - \frac{x^{1/3}}{(\tfrac{1}{3})} + 2 \frac{x^{-2}}{(-2)} - \frac{x^{-1}}{(-1)} + C$$

$$= \frac{3}{5} x^{5/3} - 3x^{1/3} - \frac{1}{x^2} + \frac{1}{x} + C.$$

4. Prove $\displaystyle \int (y - 1)^3 \frac{dy}{y} = \frac{y^3}{3} - \frac{3y^2}{2} + 3y - \log_e y + C.$

Proof.　　　　$(y - 1)^3 = y^3 - 3y^2 + 3y - 1.$

$$\therefore \; (y - 1)^3 \cdot \frac{1}{y} = \frac{(y - 1)^3}{y} = \frac{y^3 - 3y^2 + 3y - 1}{y}$$

$$= y^2 - 3y + 3 - \frac{1}{y}.$$

$$\therefore \; \int (y - 1)^3 \frac{dy}{y} = \int \left(y^2 - 3y + 3 - \frac{1}{y} \right) dy$$

$$= \int y^2 \, dy - 3 \int y \, dy + 3 \int dy - \int \frac{dy}{y}.$$

Applying formula (V) to the first two of these integrals and (XVI) to the last we get

$$\int (y - 1)^3 \frac{dy}{y} = \frac{y^3}{3} - \frac{3y^2}{2} + 3y - \log_e y + C,$$

as was to be proved.

5. Integrate $\int \dfrac{dx}{3x + 1}$

Solution. Now $d(3x + 1)$ equals $3\, dx$, so that if the numerator of the given expression were $3\, dx$ instead of dx we would have $\int \dfrac{d(3x + 1)}{(3x + 1)}$. But this requires multiplication by 3, and to compensate for this or avoid changing the value of the expression we must also divide by 3, as pointed out in the derivation of formula (IV). We can write

$$\int \frac{dx}{3x + 1} = \frac{1}{3} \int \frac{3\, dx}{3x + 1} = \frac{1}{3} \int \frac{d(3x + 1)}{3x + 1}$$

and the last expression is of the same form as $\int \dfrac{dx}{x}$. By formula (XVI), therefore,

$$\int \frac{dx}{3x + 1} = \frac{1}{3} \log_e (3x + 1) + C.$$

This completes the integration, but the result may be put into a different form. For

$$\tfrac{1}{3} \log_e (3x + 1) = \log_e (3x + 1)^{1/3} = \log_e \sqrt[3]{3x + 1}.$$

$$\therefore \quad \int \frac{dx}{3x + 1} = \log_e \sqrt[3]{3x + 1} + C.$$

6. Find $\int (\sec x - \tan x) \sec x\, dx$.

Solution. $(\sec x - \tan x) \sec x = \sec^2 x - \sec x \tan x$. Hence, the integral is

$$\int (\sec^2 x - \sec x \tan x)\, dx = \int \sec^2 x\, dx - \int \sec x \tan x\, dx,$$

and by formula (XII),

$$\int \sec^2 x\, dx = \tan x,$$

while by (XIV),

$$\int \sec x \tan x \, dx = \sec x.$$

$$\therefore \quad \int (\sec x - \tan x) \sec x \, dx = \tan x - \sec x + C.$$

7. Integrate $2(\cot \theta + \tan \theta)$ and simplify the result.

Solution. $\int [2(\cot \theta + \tan \theta)] \, d\theta = 2 \int (\cot \theta + \tan \theta) \, d\theta$

$$= 2 \int \cot \theta \, d\theta + 2 \int \tan \theta \, d\theta.$$

By formula (IX),

$$\int \cot \theta \, d\theta = \log_e (\sin \theta).$$

By formula (VIII),

$$\int \tan \theta \, d\theta = -\log_e (\cos \theta).$$

$$\therefore \quad 2 \int \cot \theta \, d\theta + 2 \int \tan \theta \, d\theta = 2 \log_e (\sin \theta) - 2 \log_e (\cos \theta)$$

$$= 2[\log_e (\sin \theta) - \log_e (\cos \theta)]$$

$$= 2 \log_e \left(\frac{\sin \theta}{\cos \theta}\right) = 2 \log_e (\tan \theta)|$$

$$= \log_e (\tan^2 \theta)$$

$$\int [2(\cot \theta + \tan \theta)] \, d\theta = \log_e (\tan^2 \theta) + C.$$

8. Find the numerical value of $\frac{1}{2} \int (\log_e x + \log_{10} x) \, dx$ when $x = 2e$.

Solution.

$$\int (\log_e x + \log_{10} x) \, dx = \int \log_e x \, dx + \int \log_{10} x \, dx.$$

By (XX),

$$\int \log_e x \, dx = x(\log_e x - 1).$$

By (XXI)

$$\int \log_{10} x \, dx = x \log_{10} \left(\frac{x}{e}\right).$$

$$\therefore \ \frac{1}{2} \int (\log_e x + \log_{10} x)\, dx = \frac{1}{2}\left[x(\log_e x - 1) + x \log_{10}\left(\frac{x}{e}\right)\right]$$

$$= \frac{x}{2}\left[\log_e x - 1 + \log_{10}\left(\frac{x}{e}\right)\right]$$

when $x = 2e$ this gives

$$\frac{2e}{2}\left[\log_e (2e) - 1 + \log_{10}\left(\frac{2e}{e}\right)\right]$$

which, on being reduced becomes $e[(\log_e 2 + \log_e e) - 1 + \log_{10} 2]$, or finally, $e(\log_e 2 + \log_{10} 2)$.

Now $e = 2.7183$, $\log_e 2 = .69315$, $\log_{10} 2 = .30103$. Using these numerical values in the last expression it becomes

$$2.7183(.69315 + .30103) = 2.7.$$

Hence when $x = 2e$,

$$\tfrac{1}{2} \int (\log_e x + \log_{10} x)\, dx = 2.7 + C.$$

9. Integrate $2.3026(10^x)$.

Solution. $\displaystyle\int 2.3026(10^x)\, dx = 2.3026 \int 10^x\, dx$ and, by formula

(XIX), $\displaystyle\int 10^x\, dx = (\log_{10} e)\cdot 10^x = .4343(10^x)$. Also 2.3026 times .4343 equals 1. Therefore,

$$\int 2.3026(10^x)\, dx = 10^x + C.$$

10. If $dW = p\, dv$ and $pv = RT$, where R and T are constants, find the value of W.

Solution. Since $pv = RT$, $p = RT/v$ and $dW = p\, dv = RT(dv/v)$.

Therefore, W which is equal to $\displaystyle\int dW$, equals

$$\int RT(dv/v) = RT \int \frac{dv}{v}.$$

$$\therefore \ \ W = RT \log_e v + C.$$

This integration and the resulting formula are of great importance in the theory of gases and steam, and in books on physics and steam engineering the constant C is determined.

Exercises.

Perform each of the following integrations after suitable algebraic transformation, when necessary, to bring the given expression into one of the standard forms of article 56.

1. $\int (5x^3 + \sqrt{x} - 2x)\, dx.$

2. $\int x(6x^2 + 7/x^4 + 4)\, dx.$ (Multiply out.)

3. $\int \frac{1}{2}(e^x - \sqrt[4]{x})\, dx.$

4. $\int \left(\frac{2}{x} - 5\sqrt[4]{x^2}\right) dx.$

5. $\int 2a \log_e x\, dx.$

6. $\int \frac{2^z\, dz}{3}.$

7. $\int \frac{\log_{10} y}{10}\, dy.$

8. $\int \frac{x^3 - x^2 - x - 1}{x^2}\, dx.$ (Simplify by division.)

9. $\int \frac{2x\, dx}{x^2 + 1}.$

10. $\int \left(\sin 3x + \cos 5x - \sin \frac{x}{2}\right) dx.$

11. $\int \frac{1 + \sin \theta}{\cos \theta}\, d\theta.$ (Express in terms of $\sec \theta$ and $\tan \theta$.)

12. $\int \sin \theta \cos \theta\, d\theta.$ (Note that $\cos \theta\, d\theta = d\,(\sin \theta)$.)

13. $\int \frac{dx}{\sqrt{5 - 4x}}.$

14. $\int \frac{4x\, dx}{\sqrt{3x}}.$

15. $\int (3 - 2x)^2 \, dx.$

16. $\int \dfrac{3x \, dx}{(x^2 + 1)^2}.$

17. $\int \sin^2 x \cdot \cos x \, dx.$

18. $\int \tan ax \cdot \sec^2 ax \, dx.$

19. $\int \left(\dfrac{\sec x}{1 + \tan x} \right)^2 dx.$

20. $\int \dfrac{(2x + 5) \, dx}{x^2 + 5x}.$

21. $\int \dfrac{2 \cos x \, dx}{4 + \sin x}.$

22. $\int (e^{\frac{x}{a}} - e^{-\frac{x}{a}})^2 \, dx.$

23. $\int xe^{(x^2)} \, dx.$

24. $\int a^x e^x \, dx.$

25. $\int \dfrac{e^{\frac{1}{x}} \, dx}{x^2}.$

Chapter 14

HOW TO USE
INTEGRAL FORMULAS

58. The Standard Forms. The integrals in the list of article 56 are the so-called standard forms, and, as we have seen in the illustrative examples, and the exercises of the last chapter, these will suffice for the integration of many expressions. Obviously, however, not all possible expressions will be of exactly these forms. It will therefore be necessary to make some transformation, or to substitute a new variable in place of the old, or to separate the given expression into parts, etc., such that it will be converted into a form which can be recognized as corresponding to one of the standard forms.

When an integral is once recognized as one of the standard forms it is at once written out by the appropriate integral formula; the first thing to be done in any case, therefore, is to examine the integral to see whether or not it is a standard form, and if it is not, the next thing to do is to transform it so as to reduce it to such a form. A large part of the work in integration, therefore, is not actually integration, but algebraic transformation and reduction, and it is evident that the student of calculus must be able to use the methods and formulas of algebra and trigonometry in order to do this work successfully and easily.

The remainder of this chapter will be concerned with methods of using the integral formulas in cases which require transformation or reduction. We shall find, however, that there are no general rules for these transformations, but each must be handled as a separate problem.

59. Algebraic Simplification. Many expressions involving products and quotients can be changed into sums and differences by carrying out the indicated multiplications and divisions. As an illustration consider the example $\int (x^2 + 3x)x \, dx$. This does not correspond to any of the standard forms, but by multiplication $(x^2 + 3x)x = x^3 + 3x^2$ and therefore $\int (x^2 + 3x)x \, dx$ becomes $\int x^3 \, dx + 3 \int x^2 \, dx = \frac{1}{4}x^4 +$

$x^3 + C.$ By the same method we can write

$$\int (ax + b)(x - c)\, dx = \int (ax^2 - acx + bx - bc)\, dx$$

$$= \int [ax^2 + (b - ac)x - bc]\, dx$$

$$= a \int x^2\, dx + (b - ac) \int x\, dx - bc \int dx$$

$$= \tfrac{1}{3}ax^3 + \tfrac{1}{2}(b - ac)x^2 - bcx + C.$$

In the integral $\int \left(\dfrac{x^3 - a^3}{x - a}\right) dx$ we can, by division, write $\dfrac{x^3 - a^3}{x - a} = x^2 + ax + a^2$ and the integral becomes

$$\int (x^2 + ax + a^2)\, dx = \int x^2\, dx + a \int x\, dx + a^2 \int dx$$

$$= \tfrac{1}{3}x^3 + \tfrac{1}{2}ax^2 + a^2 x + C.$$

In this case the quotient is exact, but in many cases it may not be so. Thus, in the case

$$\int \frac{2x^4 - 2x^3 - 4x^2 + x}{x + 3}\, dx$$

the denominator is not a factor of the numerator.   By carrying out the indicated division as far as it will go, however, and forming a fraction of the remainder, we get $2x^3 - 8x^2 + 20x - 61 + \dfrac{183}{x + 3}$, and when this is integrated term by term it gives

$$2 \int x^3\, dx - 8 \int x^2\, dx + 20 \int x\, dx - 61 \int dx + 183 \int \frac{dx}{x + 3}$$

$$= \tfrac{1}{2}x^4 - \tfrac{8}{3}x^3 + 10x^2 - 61x + 183 \log_e (x + 3) + C.$$

Generally a fraction may be reduced to a sum or difference of terms by division, if the highest power of the variable in the numerator is equal to or greater than the highest power of the variable in the denominator.

In any case of products or quotients there must be only one variable. While no general rule can be given, the above illustrations will indicate the *method* to be used in any particular example.

60. Method of Substitution.   A very useful method of integration is that of substitution of a new variable in place of some function of the given variable in the integral.   An illustration is more instructive than a description.   For example, let us find $\int \sin 3x \, dx$.   In integrating $\int \sin \theta \, d\theta = -\cos \theta$ we have $d\theta = d(\theta)$ but in our present integral dx is not equal to $d(3x)$.   So in place of $3x$ let us substitute $\theta = 3x$; then $d\theta = 3 \, dx$ and hence $dx = \frac{1}{3} \, d\theta$.   The integral then becomes

$$\int (\sin \theta)(\tfrac{1}{3} \, d\theta) = \tfrac{1}{3} \int \sin \theta \, d\theta = -\tfrac{1}{3} \cos \theta.$$

But $\theta = 3x$ and the final result is therefore $-\frac{1}{3} \cos 3x + C$, or $C - \frac{1}{3} \cos 3x$.

Similarly, in $\int \sin \theta \cos \theta \, d\theta$, let $\sin \theta = z$, then

$$dz = d(\sin \theta) = \cos \theta \, d\theta.$$

But we already have $\cos \theta \, d\theta$ in the integral.   The integral is, therefore,

$$\int (\sin \theta)(\cos \theta \, d\theta) = \int z \, dz = \tfrac{1}{2}z^2 + C.$$

But $z = \sin \theta$ and the result is therefore $\frac{1}{2} \sin^2 \theta + C$.

In general we may say that this method will work if one part of the expression to be integrated is equal to the differential of the other part or can be made so by multiplication or division by a constant.   The parts here referred to are understood to be factors.   Thus in the second case above we had

$$(\cos \theta \, d\theta) = d(\sin \theta),$$

while in the first $d(3x) = 3 \, dx$.

As a third example consider the integral $\int x\sqrt{x^2 - a^2} \, dx$.   This can be written $\int (x^2 - a^2)^{1/2} \cdot x \, dx$.   Now

$$d(x^2 - a^2) = d(x^2) - d(a^2) = 2(x \, dx),$$

therefore $x \, dx = \frac{1}{2}d(x^2 - a^2)$.   So let $x^2 - a^2 = z$ and we have

$$\int (x^2 - a^2)^{1/2} \cdot (x \, dx) = \int (z)^{1/2} \cdot \left(\frac{1}{2} dz\right) = \frac{1}{2} \int z^{1/2} \, dz = \frac{1}{2} \left(\frac{z^{3/2}}{\frac{3}{2}}\right) + C$$

or $\frac{1}{3} z^{3/2} + C$. Replacing in this the value of z the final result is

$$\tfrac{1}{3}(x^2 - a^2)^{3/2} + C, \quad \text{or} \quad \tfrac{1}{3}\sqrt{(x^2 - a^2)^3} + C.$$

The substitution used in this last example will always work in the case of the square root of a binomial in which one term is a constant and the other a power of the variable, provided the square root is multiplied by the variable raised to a power which is one less than the power under the radical.

In dealing with right triangles, or distances from the origin to points on a graph, such expressions as $a^2 - x^2$, $x^2 - a^2$, $\sqrt{x^2 \pm a^2}$, etc., are met so often that they have all been integrated and are of such frequent use that they are often included in lists of the standard forms. These may all be integrated by the method of substitution and as examples of the method we will consider a few of them here.

For example, let us integrate $\displaystyle\int \frac{dx}{x^2 + a^2}$ and use the substitution

$$x = a \tan \theta. \tag{73}$$

Then,

$$x^2 + a^2 = a^2 \tan^2 \theta + a^2 = a^2 (\tan^2 \theta + 1).$$

But, by trigonometry, $\tan^2 \theta + 1 = \sec^2 \theta$; therefore,

$$(x^2 + a^2) = a^2 \sec^2 \theta. \;\Big\}$$

Also, by (73),

$$dx = a \sec^2 \theta \, d\theta. \;\Big\}$$

Using these values the integral becomes

$$\int \frac{dx}{x^2 + a^2} = \int \frac{a \sec^2 \theta \, d\theta}{a^2 \sec^2 \theta} = \frac{1}{a} \int d\theta = \frac{1}{a} \theta + C.$$

By (73), $\tan \theta = x/a$, hence, θ is the "angle whose tangent is x/a." This is written

$$\theta = \tan^{-1}\left(\frac{x}{a}\right).$$

The result of the integration is, therefore, $\dfrac{1}{a} \tan^{-1}\left(\dfrac{x}{a}\right) + C$, or when the constant is understood to be added,

$$\int \frac{dx}{x^2 + a^2} = \frac{1}{a}\tan^{-1}\left(\frac{x}{a}\right). \qquad \text{(XXII)}$$

In order to integrate $\int \dfrac{dx}{\sqrt{x^2 + a^2}}$, let us try putting

$$\sqrt{x^2 + a^2} = z. \qquad (74)$$

Then, $x^2 + a^2 = z^2$, and differentiating this to obtain a relation involving dx, we get $2x\,dx = 2z\,dz$. Hence,

$$\frac{dx}{z} = \frac{dz}{x}.$$

This is a proportion, and, by algebra, when $a/b = c/d$ then

$$\frac{a}{b} = \frac{(a + c)}{(b + d)}.$$

Therefore, our proportion can be written $dx/z = (dx + dz)/(x + z)$. But, in this, $dx + dz = d(x + z)$. Hence,

$$\frac{dx}{z} = \frac{d(x + z)}{x + z}.$$

$$\therefore \quad \int \frac{dx}{z} = \int \frac{d(x + z)}{x + z} = \log_e (x + z) + C.$$

Replacing in the first and last member of this equation, the value of z from equation (74), we get finally

$$\int \frac{dx}{\sqrt{x^2 + a^2}} = \log_e (x + \sqrt{x^2 + a^2}) + C.$$

If the original $x^2 + a^2$ had been $x^2 - a^2$ we would have the same result with the sign changed under the radical sign. Therefore, to cover both cases, and understanding that the constant C is to be added, we can write

$$\int \frac{dx}{\sqrt{x^2 \pm a^2}} = \log_e (x + \sqrt{x^2 \pm a^2}). \qquad \text{(XXIV)}$$

These illustrations will indicate sufficiently the use of the method of substitution.

61. Integration by Parts.   In deriving the integral formulas (XX) and (XXI), article 55, we explained briefly this method of integration. As shown there, since

$$d(uv) = u \, dv + v \, du,$$

then,

$$u \, dv = d(uv) - v \, du,$$

and integrating,

$$\int u \, dv = uv - \int v \, du + C. \tag{75}$$

This formula or method may be used whenever it is required to integrate a product, one part of which is the differential of some expression, but not, however, the differential of the other factor. Thus one factor or part of the product will be the function u and the other the differential of some function v, that is, dv. Integration of the part set equal to dv, gives v, and differentiation of the part set equal to u, gives du. The resulting expressions for u, v and du are then substituted in the right side of equation (75), and if the resulting integral $\int v \, du$ can be found, the equation gives the full expression for the original integral $\int u \, dv$. If the integral $\int v \, du$ cannot be integrated, or if it gives an impossible result, the method does not work. Sometimes this second integral must itself be integrated by parts.

As illustrations of the use of this method, we already have the derivation of formulas (XX) and (XXI). As another illustration, let us integrate $\int x \sin x \cdot dx$. In this case we can easily integrate $\sin x \, dx$, so in the product $x \cdot \sin x \cdot dx$ let us put

$$u = x, \quad dv = \sin x \, dx.$$

Then

$$du = dx, \quad v = -\cos x,$$

and

$$\int u \cdot dv = u \cdot v - \int v \cdot du$$

or

$$\int (x) \cdot (\sin x \, dx) = (x) \cdot (-\cos x) - \int (-\cos x) \cdot (dx).$$

The operations indicated in the right member of this equation, when carried out, give

$$-x \cos x + \int \cos x \, dx = -x \cos x + \sin x + C.$$

Therefore,

$$\int x \sin x \, dx = \sin x - x \cos x + C.$$

As an illustration of a case in which the second integral in equation (75) must itself be integrated by parts, let us consider the integral $\int x^2 \cos x \, dx$. Put

$$u = x^2, \qquad dv = \cos x \, dx,$$

then,

$$du = 2x \, dx, \qquad v = \sin x,$$

and

$$\int u \, dv = x^2 \sin x - 2 \int x \sin x \, dx.$$

Here the last expression must itself be integrated by parts. However, in the last example, we found this equal to $\sin x - x \cos x$. Therefore, our original integral becomes

$$\int x^2 \cos x \, dx = x^2 \sin x - 2(\sin x - x \cos x)$$

$$= x^2 \sin x - 2 \sin x + 2x \cos x$$

$$= (x^2 - 2) \sin x + 2x \cos x + C.$$

It is not always an easy matter to select the proper factor for u and for dv; and sometimes either part may serve equally well for u or dv. In such cases the best choice can only be determined by trial. A case to which the method does not apply is, $\int x \tan x \, dx$. In fact this expression has never been integrated by any method.

62. Remarks. As seen in each of the examples worked out in this chapter the expression required to be integrated has in every case either been separated into parts or changed in form so that the whole expression or each of the parts separately reduces to one of the standard forms. As stated before, there are no general rules which cover all cases; each case must be handled as a separate problem, but, in general, if the expression is integrable, some such procedure will usually serve.

There are other methods applicable to other and more complicated forms of expression, but the reader must refer to more advanced books for these.

As a matter of fact, there are a very great many expressions which cannot be integrated by any methods known except by approximation. It is fortunately true, however, that most of the expressions met in ordinary engineering science can either be reduced to standard forms or else integrated by simple approximate methods as accurately as engineering measurements require.

Integration in general is as much an art as a science, and, as such, it requires a great deal of practice. The student is, therefore, urged to work all the exercises given in this and the preceding chapter and as many others as he can find in his reading.

As a test of the correctness of an integration, it is wise to differentiate the result. Differentiation should, of course, give back again the original expression.

As already mentioned, tables of integrals have been prepared which give not only the standard forms of article 56 but also many other integrals involving such combinations as those discussed in this chapter. Thus Peirce's "Table of Integrals" contains over five hundred integrals, and, in addition, lists of derivatives and other formulas useful in trigonometry and higher mathematics. The various handbooks of engineering also contain lists of the integrals which are most useful in engineering and applied science.

At the end of this book there is a table containing 63 integrals which include those most frequently useful in the ordinary applications of the calculus. If one has occasion to use integrals often this table or the larger table of Peirce mentioned above will be found very useful. In the following practice exercises, however, the reader should as far as possible use the methods of reduction explained and illustrated in this chapter together with the standard forms given in article 56.

Exercises.

Perform the integrations indicated in each of the following after a reduction by one or a combination of the methods explained in this chapter.

1. $\displaystyle\int x^2(3x^5 + 5)(3x^5 - 5)\,dx.$

2. $\displaystyle\int (2x^2 + x)^3\,dx.$

3. $\int \dfrac{x^2 + 2x + 1}{x + 1}\, dx.$

4. $\int \dfrac{x^2 - 2x + 1 + a}{x - 1}\, dx.$ (a is constant.)

5. $\int \dfrac{x^5 - 2x^4 - x^2 + 3}{x^2}\, dx.$

6. $\int \dfrac{x^2}{\sqrt{x}}\, (1 + 2x - 3x^2)\, dx.$

7. $\int \cos ax\, dx.$

8. $\int \dfrac{dx}{\sqrt{2x + 5}}.$ (Substitute $y = 2x + 5$.)

9. $\int x\sqrt{x + 3}\, dx.$ (Substitute $y = x + 3$.)

10. $\int e^{\frac{x}{2}}\, dx.$

11. $2 \int \dfrac{x + 3}{x^2 + 6x + a}\, dx.$ (Differentiate the denominator.)

12. $\int xe^x\, dx.$ (Integrate by parts; let $u = x$, $dv = e^x\, dx$.)

13. $\int x^2 e^{-x}\, dx.$ (Integrate by parts.)

14. $3 \int \dfrac{dx}{x^2 + 9}.$ (Use formula (XXII).)

15. $\int \dfrac{2\, dx}{x^2 - 1}.$ (Separate $\dfrac{2}{x^2 - 1}$ into two fractions.)

16. $\int x \log_e x\, dx.$

17. $\int \dfrac{x\, dx}{2x^2 - 3}.$

18. $2 \int \dfrac{dx}{\sqrt{x^2 - 4}}$. (Use formula (XXIV).)

19. By the substitution $x = a \sin \theta$ show that

$$\int \frac{dx}{\sqrt{a^2 - x^2}} = \sin^{-1}\left(\frac{x}{a}\right).$$ (XXIII)

20. $\int \dfrac{dx}{x^2\sqrt{x^2 - 3}}$.

21. $\int \dfrac{\sqrt{4 - x^2}}{x^2}\, dx$.

22. $\int x \cos 3x\, dx$.

23. $\int x \sec^2 2x\, dx$.

24. $\int x^2 e^{3x}\, dx$.

25. $\int \dfrac{x e^x\, dx}{(1 + x)^2}$.

Chapter 15

INTERPRETATION OF INTEGRALS BY MEANS OF GRAPHS

63. Meaning of the Integral Sign. We have given the definition of the integral sign $\int$ as simply indicating the inverse operation of differentiation. We shall now discuss more fully its real significance.

If we take the differential of a variable x we get dx. If now we integrate dx we again have x. That is,

$$\int dx = x.$$

Viewed simply as the inverse of differentiation, the integral sign is only the symbol of a certain operation to be performed on dx, and this operation is one which exactly nullifies or undoes the operation indicated by the symbol d. From this viewpoint the process of differentiation would also be simply an operation which is indicated by the symbol d and which is exactly reversed by the symbol $\int$. In this way we should simply go back and forth or in a circle and get nowhere.

We had a very clearly defined meaning for the symbol d, however, and used it in the solution of problems before we had even seen the integral sign. The quantity dx, the differential of x, is the difference in the value of x produced by the variation of the variable, the *part* of x added (or subtracted) in a process of increase (or decrease), and this part of x when divided by the interval or portion of time dt, in which the change dx occurs, gives us the *rate* of the change in x, $\dfrac{dx}{dt}$. Viewed as mathematical symbols we have found that many of the rules of algebra and arithmetic can be applied to them and certain definite things can be accomplished with them, but from the first, differentials in themselves are simply differences, *parts* of the quantities to which they refer.

145

If, then, dx is simply a part of x and by *integrating* the *parts* we get the *whole* of x, what is the integral?   It is nothing more than the *sum* of all the parts dx.   This explains the whole meaning of the symbol $\int$.

It is simply an elongated S and when placed before the symbol for differential of x or the differential of any complicated expression, it simply indicates that all the differential parts of some expression, represented by the given expression, are to be summed up to find the total or *integral* quantity.   Since summing up is simply the reverse of taking parts, it is now easy to see why it was so simple to learn to integrate by just reversing the operation of differentiation.

Now that we have studied and practiced the simple methods of integration, however, we should wish to know how to use them; how to solve problems by the aid of integrals.   Before we undertake the full interpretation of integrals of any form, let us get a graphical picture of the simple integral $\int dx = x$.

For this purpose the first diagram in the book, Fig. 1, is very suitable. Referring to Fig. 1, we recall that when distance is measured from O toward the right, the distance PP' is a part dx of the total distance x, but x is not necessarily the total length of the line, which may extend indefinitely toward the right or the left.   It is now clear at a glance that $\int dx$ is simply the *sum of all such parts as dx* and can be nothing else than the total length x as measured from O *to any chosen point* on OX.

The question now arises as to what the meaning is of such an integral as $\int x^2\, dx$.   If we let the function x^2 be represented by y, that is, $y = x^2$, then what is the interpretation of $\int y\, dx$?   Or, more generally, if $y = f(x)$, any function of x whatever, then what is the meaning of the integral $\int f(x)\, dx$ or $\int y\, dx$?   This question is answered in the next article.

64. The Area Under a Curve.   We have seen in Chapter 7 that any equation whatever, $C = \frac{1}{2}P + 5, y = x^2, y = \sin x$, or in general $y = f(x)$, where $f(x)$ is *any* function, can be plotted as a curve. We have there found that some interesting properties of curves, such as direction of the tangents, the highest points, the lowest points, the rise and fall of the curve, etc., can be determined by the aid of differentials. Such curves have also other interesting and useful properties, such as their lengths, the areas enclosed by them or included between the curve and the axes, etc.

For example, if the equation $x^2 + y^2 = a^2$, or $y = \sqrt{a^2 - x^2}$, is plotted, it will be found to give the curve of Fig. 23, which is a circle with center at the origin O and radius OM = a. From elementary geometry or mensuration in arithmetic, it is known that the length of this curve is $2\pi a$ where π is an abbreviation or symbol for the number $3.14159 + \cdots$, and that the total area enclosed by the curve is πa^2.

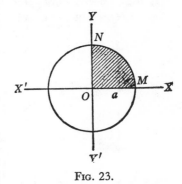

FIG. 23.

Suppose it is desired to find the area of the curve $y = x^2$, which is plotted in Fig. 15; how shall we proceed? First we must state what we mean by the "area of the curve." In the case of the circle, which is a closed curve, the area is simply that inside the curve. In Fig. 15, however, we have no closed curve; it goes on indefinitely, rising higher and higher above the axis of x and opening out farther and farther from the axis of y. To speak of the area we must mark some limits such, for example, as a line through the point P parallel to the x axis and cutting the y axis and the other branch of the curve. Then we might refer to the "area of the curve" as the area inside the lower bend of this curve below this line, or the area below this line on one side of the y axis. Not all curves, however, would be so shaped that such a line could be drawn and so such a definition for the area would not always fit.

We might also draw a line through P parallel to the y axis and cutting the x axis, and refer to the area enclosed by this line, the x axis, and the curve. This line would be the ordinate of the point P, the distance from the point O to where this line cuts the x axis would be the abscissa of P, and the area enclosed by the x axis, the ordinate and the curve would be called "the area under the curve." A little thought will make it clear that such a definition would fit any curve plotted with reference to the usual coordinate axes. Viewed in this manner, the area under the curve in Fig. 23 is the shaded area OMNO and the total area enclosed by the curve is four times this area. In general we refer to the *area under a curve* defined in this way rather than the "area of a curve" which may be an entirely different sort of thing for different curves. Let us see now how we can formulate this area in a simple manner which shall apply to any and all curves.

Let us consider the curve as having the equation $y = f(x)$, where $f(x)$ may be any expression whatever which can be plotted, and let Fig. 24 represent the curve. Then if M and N are two points on the

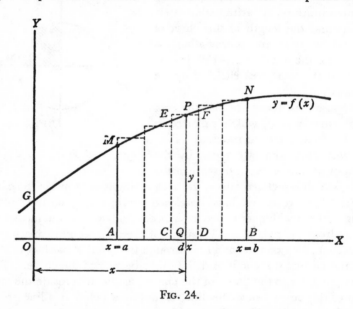

Fig. 24.

curve and AM and BN are their ordinates while P is any point on the curve with the ordinate $PQ = y$, the area under the curve will be the area ABNMA enclosed by the x axis, the ordinates at A and B, and the curve. We have now to express this area by a formula.

Since $PQ = y$ is the ordinate of P, $OQ = x$ is the abscissa, and if the length AB is divided into a number of equal parts, each will be of a length equal to a certain part of x, namely, dx. Let Q be at the center of one of these lengths CD and draw the vertical dotted lines CE and DF to meet the horizontal line EF drawn through P. Then the figure CDEF is a rectangle. Similarly, draw, or imagine drawn, other verticals at the points which divide AB into the lengths dx, and the horizontal tops of the rectangles to cut the curve as EF cuts it at P.

The curve ordinate at C is slightly less than CE, and that at D is slightly greater than DF. If CD = dx is made sufficiently small, however, PQ = y is the *average* ordinate of that section of the curve between the lines CE and DF and the area under this section of the curve will be equal to the area of the rectangle CDEF, which is equal to $\overline{EC} \cdot \overline{CD}$, or $\overline{PQ} \cdot \overline{CD}$. But $\overline{PQ} = y$ and $\overline{CD} = dx$. Therefore, the area of the small rectangle, which is equal to *the area under the small section of the curve* is $y \cdot dx$. If we let A represent the total area ABNMA, then that part of the area under the section of the curve corresponding to the length dx is dA. Therefore,

$$dA = y \, dx. \tag{76}$$

Now P is *any* point on the curve between M and N, and therefore y represents any ordinate; also all the lengths dx are equal. This expression therefore represents the area under any small section of the curve between M and N and the total area ABNMA is the sum of all such parts or differentials of the area as dA. But the sum of all the parts dA of the area is simply $\int dA = A$. By (76), therefore,

$$A = \int y \, dx. \tag{77}$$

The integral $\int y \, dx$ about which our question was asked at the end of the last article is now seen to be nothing more than *the area under a curve;* y is any ordinate of the curve and is the function $f(x)$ found by solving the equation of the curve for y.

We can now interpret the integral $\int x^2 \, dx$; it is simply the area under the curve of Fig. 15 between any two ordinates and if we take the first ordinate at the point O where the curve touches the x axis and the other at *any* point P on the curve where the ordinate is y and the abscissa is x, then the function $f(x)$ is x^2, that is, the equation of the curve is

$$y = x^2$$

and *the area under the curve* corresponding to the abscissa x is

$$A = \int y\, dx = \int x^2\, dx = \frac{x^3}{3}. \tag{78}$$

65. The Definite Integral. In the formula $\int dx = x$, when it is interpreted as the sum of all the parts of the length x along the line OX in Fig. 1, there is no indication of the actual length of the portion of OX represented by x. In order to give a definite value to the integral let us examine a little more closely the interpretation obtained in article 63. For this purpose, Fig. 1 is reproduced below as Fig. 25.

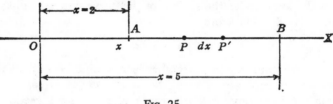

FIG. 25.

As before, x represents the length from O to any point P, and dx is the small addition PP′.

Suppose that we wish to know the length from the point A, where $x = 2$ units (centimeters, inches, feet, etc.), to the point B where $x = 5$. The length is, of course, 3 units, but how is this to be expressed by means of the integral?

In order to answer this question, let us investigate the finding of the length to be 3 in the ordinary way. If A and B are simply two different points on OX whose distances from O are not known, then we must measure AB. When we are told, however, that the distance of A is 2 and that of B is 5 from the same point O along the same line, then we simply subtract $\overline{OA} = 2$ from $\overline{OB} = 5$ and get at once $\overline{AB} = 3$. Now, the integral $\int dx = x$ gives the sum of all the differentials from O up to any point where the distance is x. If we specify then that x is equal to 2 then $\int dx = 2$, and if we specify that x is equal to 5 then $\int dx = 5$. Therefore to find the length AB we subtract the integral corresponding to $x = 2$ from the integral corresponding to $x = 5$.

This may be expressed in symbols in either of several different ways. Thus we might write

$$\overline{AB} = \int_{(\text{for } x = 5)} dx - \int_{(\text{for } x = 2)} dx$$

or,

$$\overline{AB} = \int_{(x = 5)} dx - \int_{(x = 2)} dx$$

or, more simply,

$$\overline{AB} = \int_{(x = 2)}^{(x = 5)} dx \qquad (79)$$

and either method of writing would mean the same thing, namely, that the length $\overline{AB}$ is the difference between the values of the integral corresponding to $x = 5$ and to $x = 2$. The form (79) is the more compact form and is the method generally used to indicate a definite value of the integral. It is frequently even written as

$$\overline{AB} = \int_{2}^{5} dx. \qquad (80)$$

Either of the forms (79) or (80) is allowable and both are stated in words as "the integral from $x = 2$ to $x = 5$" of dx or as "the integral between the limits 2 and 5." An integral taken between definite limits is called a *definite integral*, and the range of the values of the variable between the limits is called the *range of integration*. In (79) or (80), 2 and 5 are called, respectively, the *lower* and *upper* limits.

In order to find the numerical value of a definite integral we must, as we say in finding the length AB from Fig. 25, first integrate the expression in the usual way, then use the larger value of the variable and find the value of the integrated expression, then find its value for the smaller value of the variable, and finally subtract the second value from the first. This procedure would be indicated as follows:

$$\overline{AB} = \int_{(x = 2)}^{(x = 5)} dx = x \Big|_{(x = 2)}^{(x = 5)} = 5 - 2 = 3$$

or,

$$\overline{AB} = \int_{2}^{5} dx = x \Big|_{2}^{5} = 5 - 2 = 3$$

or,

$$\overline{AB} = \int_{2}^{5} dx = [x]_{2}^{5} = 5 - 2 = 3. \qquad (81)$$

Any of these ways of writing out the process may be used; hereafter we shall use the form (81).

As an example of the process of definite integration, let us now consider the area under the curve of Fig. 15 as given by formula (78) in the last article.   Suppose that the two ordinates are drawn, one at the point $x = 2$ and one at the point $x = 4$.   The area under the curve between these two lines, the x axis and the curve is then the integral of $y = x^2$ between the limits $x = 2$ and $x = 4$.   According to formulas (77), (78), (81), therefore, it is

$$A = \int_2^4 y\, dx = \int_2^4 x^2\, dx = \left[\frac{x^3}{3}\right]_2^4, \tag{82}$$

and finding the numerical value as explained above by putting in the values of the limits,

$$A = \frac{(4)^3}{3} - \frac{(2)^3}{3} = \frac{64}{3} - \frac{8}{3} = \frac{56}{3} = 18\tfrac{2}{3}.$$

If x is in centimeters, inches, feet, etc., A is in square cm., in., ft., etc., respectively.

To find the area under the curve in Fig. 15 from O to the ordinate at the point where $x = 2$ we integrate from zero to 2 and get

$$A' = \int_0^2 x^2\, dx = \left[\frac{x^3}{3}\right]_0^2 = \frac{(2)^3}{3} - \frac{(0)^3}{3} = \frac{8}{3}.$$

Also,

$$A'' = \int_0^4 x^2\, dx = \left[\frac{x^3}{3}\right]_0^4 = \frac{(4)^3}{3} - \frac{(0)^3}{3} = \frac{64}{3}.$$

This shows plainly that the area between $x = 2$ and $x = 4$ is

$$A = A'' - A' = \frac{64}{3} - \frac{8}{3} = 18\tfrac{2}{3}$$

so that here again we have an illustration of the meaning of the definite integral between two values of the variable, namely, the integral from the origin to the upper value minus the integral from the origin to the lower value.

We can now express definitely and concisely the area under any curve which is the graph of the equation.

$$y = f(x)$$

between the ordinates corresponding to the abscissas $x = a$ and $x = b$; it is

$$A = \int_a^b y\, dx, \tag{83}$$

the definite integral of the function between the limits a and b.   In Fig. 24 this is represented by the area ABNMA and the abscissas are $\overline{OA} = a$, $\overline{OB} = b$.

The limits a and b may have any values, zero, positive or negative. If a is zero, then A is at O and the area would be that of the figure OBNGO, Fig. 24; if $a = \overline{OA}$ and $b = \infty$ (infinity) the area is the total area between the curve and the x-axis to the right of the ordinate at A. Similarly negative values of the limits would give areas under the curve to the left of the y-axis.

If the limits are interchanged so that the original upper limit becomes the lower limit and the original lower limit becomes the upper, then instead of subtracting the value of the integral corresponding to the lower limit from that corresponding to the upper the process is reversed, and the reversal of subtraction order simply means the change of algebraic sign.   In order to interchange the limits of a definite integral, therefore, we must place a minus sign before the integral with the interchanged limits.

66. The Constant of Integration.   In expressing the length of a line or the area under a curve by the formulas

$$L = \int dx, \quad A = \int y\, dx$$

it is of course understood that the constant of integration is to be added. That is, correctly expressed,

$$L = \int dx + C, \quad A = \int y\, dx + C.$$

Then, in equation (81) we should have for the length of AB,

$$L = \int_3^5 dx = [x + C]_3^5 = (5 + C) - (3 + C),$$

$$\therefore \quad L = 5 + C - 3 - C = 5 - 3 = 2,$$

and *C disappears by cancellation.*   That is, *in the case of the definite integral there is no added constant of integration.*

This result appears again from equation (82). Here the area is

$$A = \int_2^4 x^2\, dx = \left[\frac{x^3}{3} + C\right]_2^4 = \left[\frac{(4)^3}{3} + C\right] - \left[\frac{(2)^3}{3} + C\right],$$

$$\therefore \quad A = \tfrac{64}{3} + C - \tfrac{8}{3} - C = \tfrac{64}{3} - \tfrac{8}{3} = 18\tfrac{2}{3}.$$

Thus in definite integration the integration constant is not written after the value of the integral. To distinguish between the two methods of integration,

$$\int dx = x + C,$$

$$\int_a^b dx = [x]_a^b = b - a,$$

the first is called the *indefinite integral*, the second being the *definite integral from a to b*.

In some problems the value of the constant of indefinite integration is determined by the data or the conditions of the problem. Cases of this kind will appear in the illustrative examples below and in the next chapter.

67. The Length of a Curve. In this article we shall consider a matter which is more an application of integration than a part of the interpretation of integration, but since it is an application to the calculations of graphs we shall include it in the present chapter.

If a graph as plotted forms a loop or a closed curve, then, of course, the total length of the curve is as definite as the area. If the curve is not closed, but goes on indefinitely, then the length is as indefinite as the area. By using the same method as in describing the area, however, and stating definite limits between which the length is to be measured, the length is perfectly definite for ordinary graphs. We shall see now that it can be found by means of a definite integral when the limits are specified. If the limits are not specified, the length can still be expressed by means of an integral, but it will in this case be an indefinite integral. It is obvious that in either case, whether we can find a definite numerical value or not, the length will be the integral sum of the differentials of length along the curve.

Let us refer back to Figs. 2 and 9. We saw there that the differential of length *ds* of a curve is the distance a moving point would pass over in the interval of time *dt if the point should continue to move in the direction in which it was moving at the beginning of the interval dt.* The differential length *ds* at any point on the curve is measured instantaneously *along the tangent* at that point. As the tangent changes its direction continually to *follow the curve*, however, the sum of all the length differentials *ds* is the total length *s* along the curve.

If, therefore, the differential *ds* is taken short enough, the tangent will in that short length coincide with the curve and *ds* can be represented as in Fig. 26. The curve MN represents the graph of any

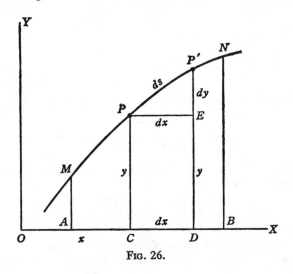

FIG. 26.

equation $y = f(x)$, and CP $= y$ is the ordinate of any point P on the curve having the abscissa OC $= x$. Then, CD $=$ PE $= dx$ is the very small differential of *x*, and EP$'$ $= dy$ the corresponding very small differential of *y* resulting from the passage of the point P to the very near point P$'$ over the very small differential of curve length PP$'$ $= ds$.

If we can get an expression for the differential length *ds* in terms of *x* and *y* from the equation of the curve, then the length of the curve will be the sum of all the parts *ds*, that is, the integral of the formula for *ds*. If we specify two points such as M and N, then the length MPN will be given by the definite integral from $x =$ OA to $x =$ OB.

To find an expression for *ds*, we have from the small right triangle PEP$'$, $\overline{PP'}^2 = \overline{PE}^2 + \overline{EP'}^2$ or,

$$(ds)^2 = (dx)^2 + (dy)^2.$$

If we divide both sides of this equation by $(dx)^2$, we get

$$\frac{(ds)^2}{(dx)^2} = 1 + \frac{(dy)^2}{(dx)^2}$$

or,

$$\left(\frac{ds}{dx}\right)^2 = 1 + \left(\frac{dy}{dx}\right)^2.$$

By taking the square root of both sides of this equation,

$$\frac{ds}{dx} = \sqrt{1 + \left(\frac{dy}{dx}\right)^2},$$

$$\therefore \quad ds = \sqrt{1 + \left(\frac{dy}{dx}\right)^2} \cdot dx. \tag{84}$$

This is the desired formula for ds.   Now the total length is $s = \int ds$,

$$\therefore \quad s = \int \sqrt{1 + \left(\frac{dy}{dx}\right)^2} \cdot dx + C. \tag{85}$$

If in Fig. 26 OA $= a$ and OB $= b$, then the length between the two points M and N on the curve is

$$s = \int_a^b \sqrt{1 + \left(\frac{dy}{dx}\right)^2} \cdot dx, \tag{86}$$

the constant C not appearing in the definite integral.

The formulas (85) and (86) mean that, from the equation of the curve

$$y = f(x),$$

we are to find the derivative of y with respect to x,

$$\frac{dy}{dx} = f'(x),$$

and this expression is then to be squared and added to 1 under the radical sign.   The final expression is then to be integrated.   If no definite limit points are given, the formula (85) will give a general formula for s involving the indefinite constant C.   If limit points on the curve are specified by giving their abscissas $x = a$ and $x = b$, then formula (86) will give the definite length of the curve between these points.

As an example, let us use formula (86) to find the length of the curve of Fig. 15 between the two points whose abscissas are $x = 0$ and $x = 4$. The equation of the curve is $y = x^2$. Hence

$$\frac{dy}{dx} = 2x \quad \text{and} \quad \left(\frac{dy}{dx}\right)^2 = 4x^2,$$

and

$$s = \int_0^4 \sqrt{1 + 4x^2} \, dx. \tag{87}$$

Now we can write $\sqrt{1 + 4x^2} = \sqrt{4x^2 + 1} = \sqrt{4(x^2 + \frac{1}{4})}$ and since $\frac{1}{4} = (\frac{1}{2})^2$ we get finally $\sqrt{1 + 4x^2} = 2\sqrt{x^2 + (\frac{1}{2})^2}$ and this in (87) gives

$$s = 2 \int_0^4 \sqrt{x^2 + (\tfrac{1}{2})^2} \, dx. \tag{88}$$

If, therefore, we can integrate $\int \sqrt{x^2 + (\frac{1}{2})^2} \, dx$, the length of the specified portion of the curve is known as soon as the limits are substituted in the result of the integration and s calculated.

In problem 4, article 72, we shall find that

$$\int \sqrt{x^2 + a^2} \, dx = \tfrac{1}{2}[x\sqrt{x^2 + a^2} + a^2 \log_e (x + \sqrt{x^2 + a^2})]$$

and our integral in (88) is the same with $a = \frac{1}{2}$. Therefore (since the 2 before the integral cancels the $\frac{1}{2}$ before the bracket in the result)

$$s = 2 \int_0^4 \sqrt{x^2 + (\tfrac{1}{2})^2} \, dx$$

$$= [x\sqrt{x^2 + (\tfrac{1}{2})^2} + (\tfrac{1}{2})^2 \log_e (x + \sqrt{x^2 + (\tfrac{1}{2})^2})]_0^4.$$

If, now, the calculation of the definite integral is carried out by substituting the 4 and the 0 for x in the bracketed expression and carrying out the indicated squarings, root extractions and taking of logarithms, and the two values subtracted, as described in article 65, we get finally for the length between the given points

$$s = 16.82$$

and the units are the same as those in which x, y are measured.

68. Remarks. Although we have discussed the meaning of integration and defined the definite integral by reference to graphs and while many useful applications of integration are found in the study of graphs, it must not be supposed that the only sums of differential parts which can be interpreted as integrals are the sums of differentials of straight lines, curves or areas.   Any quantity which is susceptible of measurement may be expressed by means of symbols and when a functional relation is once expressed between *any* two such quantities we may differentiate or integrate it.   Since such a relation may always be plotted in a graph, however, we may take the graphical interpretation as correctly representing the meaning of integration in general by means of a sum.

Thus there may be differentials of volume, mass, force, work, velocity, length, angle, electric current, electromotive force, simple number, value, etc., and we have already seen in Chapters 4 and 9 that they may be used to solve problems involving the variation of such quantities.   We shall also find, in Chapter 17, that when the nature of a problem is such that we can formulate the relation between small parts of such quantities or between their rates of change, the process of integration viewed as a summing process will give us the relation between the quantities themselves.

Integration in any connection is then to be viewed in its true aspect as the summing of differentials, and the process considered and first defined as anti-differentiation refers simply to the *method of performing* the summation, that is, the mathematical operation.

The examples and exercises in this chapter will only have to do with practice in the calculation of definite integrals as interpreted in the preceding articles.   The uses of definite integrals in solving problems will be given in the next two chapters.

69. Illustrative Examples of Definite Integrals.   In the solved examples of this article, and in the unsolved exercises which follow, it is to be remembered that the expression is to be integrated as given without regard to the limits, and the integrated expression written out with the limits indicated and without the integration constant.   The value of the expression is then to be calculated with the upper limit substituted for the variable, and from this value is to be subtracted the value of the same expression found by substituting the lower limit for the variable.

1. Find the value of $\int_2^3 (x^2 - 4)^2 x \, dx$.

Solution. $(x^2 - 4)^2 x = (x^4 - 8x^2 + 16)x = x^5 - 8x^3 + 16x$

$$\int (x^5 - 8x^3 + 16x)\, dx = \int x^5\, dx - 8 \int x^3\, dx + 16 \int x\, dx$$

$$= \frac{x^6}{6} - 2x^4 + 8x^2 = x^2 \left(\frac{x^4}{6} - 2x^2 + 8 \right)$$

$$\therefore \quad \int_2^3 (x^2 - 4)^2 x\, dx = \left[x^2 \left(\frac{x^4}{6} - 2x^2 + 8 \right) \right]_2^3$$

$$= \left[3^2 \left(\frac{3^4}{6} - 2 \cdot 3^2 + 8 \right) \right]$$

$$- \left[2^2 \left(\frac{2^4}{6} - 2 \cdot 2^2 + 8 \right) \right]$$

$$= \left[9 \times \frac{7}{2} \right] - \left[4 \times \frac{8}{3} \right] = \frac{63}{2} - \frac{32}{3} = 20\frac{5}{6}.$$

2. Find the value of $\displaystyle\int_{a/2}^a \frac{dy}{\sqrt{a^2 - y^2}}.$

Solution. By formula (XXIII), article 62,

$$\int \frac{dy}{\sqrt{a^2 - y^2}} = \sin^{-1} \left(\frac{y}{a} \right)$$

$$\int_{a/2}^a \frac{dy}{\sqrt{a^2 - y^2}} = \left[\sin^{-1} \left(\frac{y}{a} \right) \right]_{a/2}^a = \sin^{-1} \left(\frac{a}{a} \right) - \sin^{-1} \left(\frac{a/2}{a} \right)$$

$$= \sin^{-1} 1 - \sin^{-1} \frac{1}{2} = \frac{\pi}{2} - \frac{\pi}{6} = \frac{\pi}{3}.$$

3. Find the value of $\displaystyle\int_5^9 \frac{x\, dx}{\sqrt{x^2 + 144}}.$

Solution. Using the method of substitution let $z = x^2 + 144$. Then $dz = 2x\, dx$ or $x\, dx = \frac{1}{2}\, dz$ and hence

$$\int \frac{x\, dx}{\sqrt{x^2 + 144}} = \int \frac{\frac{1}{2}\, dz}{\sqrt{z}} = \frac{1}{2} \int z^{-1/2}\, dz$$

$$= \frac{1}{2} \left(\frac{z^{1/2}}{\frac{1}{2}} \right) = \sqrt{z} = \sqrt{x^2 + 144}.$$

$$\therefore \quad \int_5^9 \frac{x\, dx}{\sqrt{x^2 + 144}} = [\sqrt{x^2 + 144}]_5^9 = \sqrt{9^2 + 144} - \sqrt{5^2 + 144}$$

$$= \sqrt{225} - \sqrt{169} = 15 - 13 = 2.$$

4. Find the value of $\int_{\pi/6}^{\pi/2} \cos \theta \, d\theta$.

Solution. $\int_{\pi/6}^{\pi/2} \cos \theta \, d\theta = [\sin \theta]_{\pi/6}^{\pi/2} = \sin \dfrac{\pi}{2} - \sin \dfrac{\pi}{6}$

$$= \sin 90° - \sin 30°$$

$$= 1 - \tfrac{1}{2} = \tfrac{1}{2}.$$

5. Find the value of $\int_{e}^{10e} \log_{10} x \, dx$.

Solution. By formula (XXI), Chapter 13,

$$\int \log_b x \, dx = x \log_b \left(\frac{x}{e}\right)$$

and in the present case the base $b = 10$. Therefore

$$\int_{e}^{10e} \log_{10} x \, dx = \left[x \log_{10} \left(\frac{x}{e}\right) \right]_{e}^{10e} = 10e \log_{10} \left(\frac{10e}{e}\right) - e \log_{10} \left(\frac{e}{e}\right)$$

$$= 10e \log_{10} 10 - e \log_{10} 1$$

$$= 10e \times 1 - e \times 0 = 10e = 27.183.$$

6. Find the value of $\int_{\pi/2}^{\pi} x \sin x \, dx$.

Solution. By the next to last illustration in article 61

$$\int x \sin x \, dx = \sin x - x \cos x.$$

Therefore,

$$\int_{\pi/2}^{\pi} x \sin x \, dx = [\sin x - x \cos x]_{\pi/2}^{\pi}$$

$$= (\sin \pi - \pi \cos \pi) - \left(\sin \frac{\pi}{2} - \frac{\pi}{2} \cos \frac{\pi}{2}\right)$$

$$= (\sin 180° - \pi \cos 180°) - \left(\sin 90° - \frac{\pi}{2} \cos 90°\right)$$

$$= [0 - \pi(-1)] - \left[1 - \frac{\pi}{2}(0)\right] = \pi - 1 = 2.1416.$$

Exercises.

Find the numerical value of each of the following definite integrals:

1. $\int_1^2 (x + 1)\, dx.$

2. $\int_0^2 (x + 1)(x^2 - 3)\, dx.$

3. $\int_0^\pi \sin x\, dx.$

4. $\int_1^{100} \log_{10} x\, dx.$

5. $\int_a^{2a} \left(x^3 - \dfrac{1}{\sqrt[3]{x}} \right) dx.$

6. $\int_0^{10} \dfrac{x^3}{3}\, dx.$

7. $\int_{10}^0 \dfrac{x^3}{3}\, dx.$

8. $\int_1^2 \dfrac{dy}{y}.$

9. $\int_0^5 (6 + 8x - 3x^2)\, dx.$

10. $-\int_0^\pi \cos \theta \sin \theta\, d\theta.$

11. $\int_0^2 \tfrac{1}{2}(e^x + e^{-x})\, dx.$

12. $\int_0^2 \tfrac{1}{2}(e^x - e^{-x})\, dx.$

13. $\int_0^{\pi/4} \tan x\, dx.$

14. $\int_0^{\pi/2} 3 \sin^2 \theta \cos \theta\, d\theta.$

15. $\int_e^1 (\log_e x - 1)\, dx.$

16. $\int_2^6 \sqrt{x - 2}\, dx.$

17. $\int_0^4 x\sqrt{x^2 + 9}\, dx.$

18. $\int_0^x (\sqrt{a} - \sqrt{x})^2\, dx.$

19. $\int_0^1 te^t\, dt.$

20. $\int_0^\pi e^\theta \sin \theta\, d\theta.$

Chapter 16

GRAPHICAL APPLICATIONS
OF INTEGRATION

A. AREAS UNDER CURVES

70. Illustrative Problems.   1. Find the area of a circle of radius *a*.
Solution 1.   The equation of a circle of radius *a* and with its center at the origin O, Fig. 27, is $x^2 + y^2 = a^2$, or

$$y = \sqrt{a^2 - x^2}. \tag{a}$$

The area of the shaded section OAB in Fig. 27 is the area given by the integral $\int y \, dx$ taken from $x = 0$ to $x = OA = a$, where y is given by equation (a), the equation of the curve.   The shaded area is one quadrant of the circle, and, therefore, the total area is $4 \int y \, dx$.

$$\therefore \quad A = 4 \int_0^a \sqrt{a^2 - x^2} \, dx. \tag{b}$$

The integration of $\int \sqrt{a^2 - x^2} \, dx$ is carried out below in problem 7; the result is $\dfrac{1}{2} \left[x\sqrt{a^2 - x^2} + a^2 \sin^{-1} \left(\dfrac{x}{a} \right) \right]$.

$$\therefore \quad A = 4 \left\{ \frac{1}{2} \left[x\sqrt{a^2 - x^2} + a^2 \sin^{-1} \left(\frac{x}{a} \right) \right]_0^a \right\}$$

$$= 2 \left[a\sqrt{a^2 - a^2} + a^2 \sin^{-1} \left(\frac{a}{a} \right) \right]$$

$$- 2 \left[0\sqrt{a^2 - 0^3} + a^2 \sin^{-1} \left(\frac{0}{a} \right) \right]$$

$$= 2[0 + a^2 \sin^{-1} 1] - 2[0 + a^2 \sin^{-1} 0]$$

$$= 2a^2 \sin^{-1} 1 - 2a^2 \sin^{-1} 0.$$

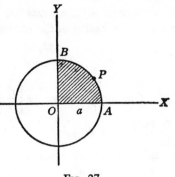

Fig. 27.

Now, $\sin^{-1} 1 = 90° = \dfrac{\pi}{2}$ and $\sin^{-1} 0 = 0°$, therefore,

$$A = 2\left(a^2 \cdot \frac{\pi}{2}\right) = \pi a^2$$

which is the familiar formula for the area of a circle of radius a.

Solution 2. Let dr be the width of a very narrow ring (shown dotted in Fig. 28), with center at the center of the circle and middle at a dis-

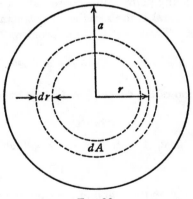

Fig. 28.

tance r from the center. This is a part dA of the area of the circle A and as in problem 3, article 21,

$$dA = 2\pi r \, dr.$$

The total area is the sum of all the parts dA, corresponding to all values of r from zero to a, the radius of the circle. Therefore,

$$A = \int_{r=0}^{r=a} dA = \int_0^a 2\pi r \, dr.$$

$$\therefore \quad A = 2\pi \int_0^a r \, dr = 2\pi \left[\frac{r^2}{2}\right]_0^a,$$

or,

$$A = \pi a^2.$$

2. Find the area under the curve whose equation is $y^2 = 4ax$, between the origin and the ordinate at $x = a$.

Solution. From the equation of the curve, $y = 2\sqrt{ax} = 2\sqrt{a} \cdot x^{1/2}$. Therefore

$$A = \int_0^a y \, dx = 2\sqrt{a} \int_0^a x^{1/2} \, dx = 2\sqrt{a} \left[\frac{x^{3/2}}{\frac{3}{2}}\right]_0^a = \frac{4\sqrt{a}}{3} [x^{3/2}]_0^a$$

$$= \frac{4\sqrt{a}}{3} [a^{3/2} - 0^{3/2}] = \frac{4a^{1/2}}{3} \cdot a^{3/2} = \tfrac{4}{3} a^{4/2}.$$

$$\therefore \quad A = \tfrac{4}{3}a^2.$$

3. Find the area between the curve of problem 2 above, the x axis, and the ordinate to any point whose coordinates are x, y.

Solution. The integral is the same as in problem 2 with the upper limit x instead of a. Therefore by (a) of problem 2,

$$A = 2\sqrt{a} \int_0^x x^{1/2} \, dx = \tfrac{4}{3}a^{1/2}x^{3/2}. \tag{b}$$

From the equation of the curve $y^2 = 4ax$, we have, as above, at the point $P(x, y)$

$$y = 2a^{1/2}x^{1/2}. \tag{c}$$

Now from (b) we have, $A = \tfrac{4}{3}a^{1/2} \cdot x \cdot x^{1/2} = \tfrac{2}{3}x(2a^{1/2}x^{1/2})$, and, hence, by (c),

$$A = \tfrac{2}{3}xy. \tag{d}$$

In Fig. 29, which gives the graph of the equation $y^2 = 4ax$, the area of the rectangle OAPB is xy. According to the result (d), therefore, the corresponding area under the curve is two-thirds that of the rectangle. This curve is called the *parabola* and the result here obtained is one of many interesting properties which this curve possesses. We shall later investigate some of its other properties (article 86).

4. Find the area of a sector of the curve $x^2 - y^2 = a^2$ included between the curve, the x axis and a line drawn from the origin to a point having the coordinates x, y.

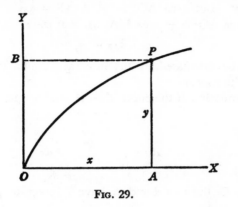

FIG. 29.

Solution. That portion of the graph of this equation which is included in the first coordinate quadrant is plotted in Fig. 30 as MN. P is the point (x, y), OA $= x$, AP $= y$, and OM $= a$ is constant. OP is the line referred to and the shaded sector OMPO is the area to be found. Let this area be denoted by u.

Now the integral formula for area,

$$A = \int y \, dx \tag{a}$$

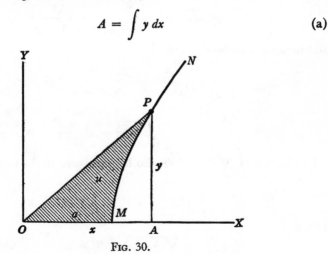

FIG. 30.

gives the area *under the curve*, that is, the area between the curve, the x axis and the ordinate y, which in Fig. 30 is the area MAPM. The area u is the difference between this area and the area of the right triangle AOP. But area AOP = $\frac{1}{2}\overline{OA}\cdot\overline{AP}$ = $\frac{1}{2}xy$. Therefore area OMPO = area AOPA − area MAPM, that is,

$$u = \tfrac{1}{2}xy - A, \tag{b}$$

and A is to be calculated by the integral formula (a) above. We proceed now to find A.

From the equation of the curve $x^2 - y^2 = a^2$ we get

$$y = \sqrt{x^2 - a^2}.$$

$$\therefore \quad A = \int_0^x \sqrt{x^2 - a^2}\, dx \tag{c}$$

is the formula for the area A and we have to integrate $\int \sqrt{x^2 - a^2}\, dx$.

This integral cannot be handled directly, but by a simple transformation it can be replaced by a form which we have already integrated.

Now $(\sqrt{x^2 - a^2})(\sqrt{x^2 - a^2}) = x^2 - a^2$ by definition of a square root. Therefore $\sqrt{x^2 - a^2} = \dfrac{x^2 - a^2}{\sqrt{x^2 - a^2}}$ and by dividing both terms in the numerator of the fraction this gives

$$\sqrt{x^2 - a^2} = \frac{x^2}{\sqrt{x^2 - a^2}} - \frac{a^2}{\sqrt{x^2 - a^2}},$$

and hence the integral becomes

$$\int \left(\frac{x^2}{\sqrt{x^2 - a^2}} - \frac{a^2}{\sqrt{x^2 - a^2}} \right) dx.$$

$$\therefore \quad \int \sqrt{x^2 - a^2}\, dx = \int \frac{x^2\, dx}{\sqrt{x^2 - a^2}} - a^2 \int \frac{dx}{\sqrt{x^2 - a^2}}. \tag{d}$$

Integrating $\int \sqrt{x^2 - a^2}\, dx$ also by parts and in the formula

$$\int U\, dV = UV - \int V\, dU$$

letting

$$U = \sqrt{x^2 - a^2}, \quad dV = dx,$$

we get

$$dU = \frac{x\,dx}{\sqrt{x^2 - a^2}}, \qquad V = x.$$

$$\therefore \quad UV = x\sqrt{x^2 - a^2}, \quad V\,dU = \frac{x^2\,dx}{\sqrt{x^2 - a^2}}.$$

$$\therefore \quad \int \sqrt{x^2 - a^2}\,dx = x\sqrt{x^2 - a^2} - \int \frac{x^2\,dx}{\sqrt{x^2 - a^2}}.$$

Also by (d),

$$\int \sqrt{x^2 - a^2}\,dx = -a^2 \int \frac{dx}{\sqrt{x^2 - a^2}} + \int \frac{x^2\,dx}{\sqrt{x^2 - a^2}}.$$

If we add these two equations the last terms cancel and we get

$$2 \int \sqrt{x^2 - a^2}\,dx = x\sqrt{x^2 - a^2} - a^2 \int \frac{dx}{\sqrt{x^2 - a^2}}.$$

$$\therefore \quad \int_a^x \sqrt{x^2 - a^2}\,dx = \tfrac{1}{2}x\sqrt{x^2 - a^2} - \tfrac{1}{2}a^2 \int_a^x \frac{dx}{\sqrt{x^2 - a^2}}. \qquad \text{(e)}$$

But $\int_a^x \sqrt{x^2 - a^2}\,dx = A$ and $\sqrt{x^2 - a^2} = y$. The equation (e), therefore, is

$$A = \tfrac{1}{2}xy - \tfrac{1}{2}a^2 \int_a^x \frac{dx}{\sqrt{x^2 - a^2}}.$$

If now we substitute this value of A in equation (b) the $\tfrac{1}{2}xy$ terms cancel and we get

$$u = \frac{a^2}{2} \int_a^x \frac{dx}{\sqrt{x^2 - a^2}}. \qquad \text{(f)}$$

Instead of the integral $\int \sqrt{x^2 - a^2}\,dx$ of equation (c) we now have $\int \frac{dx}{\sqrt{x^2 - a^2}}$ and this is one of the forms we have already integrated. (See formula (XXIV), article 60.)   It is equal to $\log_e (x + \sqrt{x^2 - a^2})$. Using this in equation (f) the desired area is

$$u = \frac{a^2}{2} \left[\log_e (x + \sqrt{x^2 - a^2}) \right]_a^x$$

$$= \frac{a^2}{2} \left[\log_e (x + \sqrt{x^2 - a^2}) - \log_e (a + \sqrt{a^2 - a^2}) \right]$$

$$= \frac{a^2}{2} \left[\log_e (x + \sqrt{x^2 - a^2}) - \log_e a \right],$$

and since the difference between two logarithms is the logarithm of a quotient, this becomes

$$u = \frac{a^2}{2} \log_e \left(\frac{x + \sqrt{x^2 - a^2}}{a} \right).$$

But $\sqrt{x^2 - a^2} = y$. Therefore, we get finally for the required area u in Fig. 30,

$$u = \frac{a^2}{2} \log_e \left(\frac{x + y}{a} \right). \tag{g}$$

If in equation (e) we substitute the value of the integral on the right side as given by formula (XXIV) and use either the plus or minus sign under the square root sign we get the formula

$$\int \sqrt{x^2 \pm a^2}\, dx = \tfrac{1}{2}[x\sqrt{x^2 \pm a^2} + a^2 \log_e (x + \sqrt{x^2 \pm a^2})], \tag{XXV}$$

which will be useful later.

5. Show by integration that the area of a right triangle of legs a, b is $\tfrac{1}{2}ab$.

Solution. Let the triangle be OAB in Fig. 31 and take OA as x-axis with OY parallel to AB. Then OA $= a$, AB $= b$, and at any point P on the line OB let the coordinates be x, y. By similar triangles we then have $y:x::b:a$. Hence

$$y = \frac{b}{a} x.$$

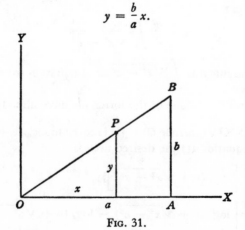

FIG. 31.

This is the equation of the line OB, since it gives the relation between the coordinates for any point on it.

The area of the triangle OAB is the area under the line OB from O to the ordinate AB, that is $A = \int_0^a y\,dx$. Hence

$$A = \frac{b}{a} \int_0^a x\,dx = \frac{b}{a} \left[\frac{x^2}{2} \right]_0^a = \frac{b}{a} \left[\frac{a^2}{2} - \frac{0^2}{2} \right] = \frac{b}{a} \cdot \frac{a^2}{2}.$$

$$\therefore \quad A = \tfrac{1}{2}ab.$$

6. A small weight lying on a flat surface is attached to the end of a string of a fixed length and the other end of the string is moved along a straight line at right angles to the original position of the string. Find the total area between the paths of the two ends of the string.

Solution. Let a be the length of the string and, in Fig. 32, let B be

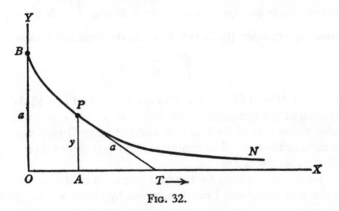

Fig. 32.

the original position of the weight and OB that of the string. Then, as the end O is moved along the line OX, the weight will slide along the curve BPN, constantly approaching OX.

When the end O of the string has reached any position T on OX the sliding weight will occupy the position P, and since PT represents the momentary direction of motion of P (the direction of tension in the string PT) then PT is the tangent to the curve BPN at the point P. Therefore the angle PTO is the angle whose tangent is the slope of the curve at P and hence

$$\frac{dy}{dx} = -\tan \text{PTO}, \tag{a}$$

since the slope is negative. Now $AP = y$ is the ordinate of P and in the right triangle PTA, $\overline{AT} = \sqrt{a^2 - y^2}$. Therefore

$$\tan PTO = \overline{AP}/\overline{AT} = y/(\sqrt{a^2 - y^2}).$$

Hence by (a),

$$\frac{dy}{dx} = -\frac{y}{\sqrt{a^2 - y^2}}.$$

$$\therefore \quad y\, dx = -\sqrt{a^2 - y^2}\, dy. \tag{b}$$

The area between the curve and the line OX is $\int y\, dx$ and since at the beginning of the motion $y = OB = a$ and ultimately P will reach OX and make $y = 0$, the area is $\int_{y=a}^{y=0} y\, dx$. If we interchange the limits we reverse the sign of the integral, that is $\int_a^0 y\, dx = -\int_0^a y\, dx$. Substituting equation (b) in this formula the required area is

$$A' = \int_0^a \sqrt{a^2 - y^2}\, dy. \tag{c}$$

The curve BPN of Fig. 32 is called the *tractrix* and Fig. 32 shows only one branch of the complete curve, since the weight may be originally taken on either side of the line OX and the point O may move either to the right or the left.   The complete curve is therefore a curve of four branches as shown in Fig. 33.

The total area enclosed by the four branches of the tractrix is four times the area under one branch as given by formula (c).   This total area is

$$A = 4 \int_0^a \sqrt{a^2 - y^2}\, dy. \tag{d}$$

Since this formula is the same as formula (b) of Problem 1 above, the total area of the tractrix is the same as that of the circle of radius a, namely, that given by the formula

$$A = \pi a^2,$$

when the length of the generating string is the same as the radius of the circle.

The area of the tractrix was first found by Sir Isaac Newton, inventor of the calculus, and it is said to be the first curve whose area was found by integration.

Note. We have thus far used integrals to find areas.   In the follow-

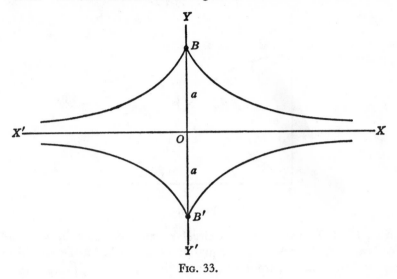

FIG. 33.

ing problem we reverse the process and use an area to find the value of an integral.

7. Find the integral $\int \sqrt{a^2 - x^2}\, dx$ by means of the graph of the equation $y = \sqrt{a^2 - x^2}$ or $x^2 + y^2 = a^2$.

Solution.   The graph of the equation $x^2 + y^2 = a^2$ is the circle with center at O and radius equal to a.   One fourth of the curve is drawn in Fig. 34 as the curve MPB.   If P represents any point on the curve with coordinates x, y, then, in the right triangle OAP,

$$y = \sqrt{a^2 - x^2} \tag{a}$$

and the given integral $\int y\, dx$ is the area under the curve from O to A, that is, the area OAPBO.   If we let A represent this area, then

$$\int y\, dx = \int \sqrt{a^2 - x^2}\, dx = A \tag{b}$$

and

$$A = \text{area OAPBO} = (\text{OAPO}) + (\text{OPBO}). \tag{c}$$

Now, (OAPO) is the right triangle of base OA = x and altitude AP = y and the area is $\frac{1}{2}$ (base $\times$ altitude). Hence

$$(OAPO) = \tfrac{1}{2}xy. \tag{d}$$

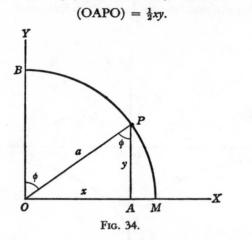

Fig. 34.

Also, (OPBO) is the area of a circular sector of radius OP = a and central angle POB = ϕ. By equation (29), article 23, the arc PB = $a\phi$, and the area is $\frac{1}{2}$ (radius $\times$ arc). Hence,

$$(OPBO) = \tfrac{1}{2}a^2\phi. \tag{e}$$

Using (d) and (e) in (c), the desired area is,

$$A = \tfrac{1}{2}(xy + a^2\phi). \tag{f}$$

In the right triangle OAP, angle APO = angle POB = ϕ, and $\sin\phi = \overline{OA}/\overline{OP} = x/a$, or $\phi = \sin^{-1}(x/a)$. Using this value of ϕ and the value of y given by (a), in the formula (f) we get

$$A = \frac{1}{2}\left[x\sqrt{a^2 - x^2} + a^2 \sin^{-1}\left(\frac{x}{a}\right) \right], \tag{g}$$

and since this is the area expressed by formula (b) it is the value of the given integral.

If some other ordinate instead of OB were used as the initial ordinate or starting point, the value of x would be the same, OA, but a constant area C not depending on x would have to be added to A as given by (g). This is the integration constant which must be added to the value of the integral as given by (g). Writing the complete integral, with this constant understood as being added, we have finally as the result of the integration, by (b) and (g),

$$\int \sqrt{a^2 - x^2}\, dx = \frac{1}{2}\left[x\sqrt{a^2 - x^2} + a^2 \sin^{-1}\left(\frac{x}{a}\right)\right]. \quad \text{(XXVI)}$$

71. Problems for Solution.

Find the area under the graph of each of the following equations between the ordinates corresponding to the two values of x given in parentheses for each:

1. $y = x^2 - 6x + 5$ $(x = 1, 4)$
2. $y = \sin x$ $(x = 0°, 180°)$
3. $y^2 = 4(x^2 - x^4)$ $(x = 0, 1)$
4. $y = 4/x^2$ $(x = 1, 3)$
5. $y = e^{-\frac{x}{4}}$ $(x = 0, 4)$
6. $y = e^{-\frac{x}{4}} + \sin x$ $(x = 0, \pi)$

7. A straight road crosses a narrow river at two points half a mile apart and the bend of the river takes such a course between the crossings that at any point on the river bank the distance to the nearest point on the road is equal to the distance of that point on the road from the first crossing minus twice the square of that distance.   What is the value of the land between the road and the river at fifty dollars an acre?

(*Hint:* Take the road as OX and the first crossing as O.   One square mile is 640 acres.)

8. Plot the graph of the equation $y = \log_e x$, find where it crosses the x-axis, and find the area under the curve between this point and the ordinate corresponding to the abscissa $x = e$.

9. The equation of the circle $x^2 + y^2 = a^2$ can be written as $\dfrac{x^2}{a^2} + \dfrac{y^2}{a^2} = 1$ and the area πa^2 as $\pi a \cdot a$.   The graph of the equation is shown in Fig. 35(a). In Fig. 35(b) is shown the graph of the equation $\dfrac{x^2}{a^2} + \dfrac{y^2}{b^2} = 1$, which is called

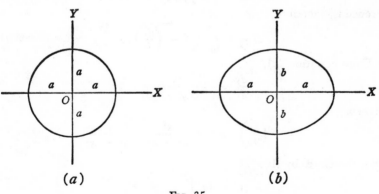

(*a*)   (*b*)

Fig. 35.

an *ellipse*. Show by integration as in Problem 1 of the preceding article that the area of this curve is $\pi a \cdot b$.

10. The graph of the equation $y = 8a^3/(x^2 + 4a^2)$, in which a is constant, was discovered by an Italian woman mathematician, Maria Agnesi, and is called the "witch." It is symmetrical with respect to the OY axis as a center line and extends to infinity in each direction along OX. Show that the area between the curve and the X axis is equal to the surface area of a sphere of radius a.

11. Using a convenient scale of length to represent the constant a plot the "witch" of Problem 10 and the curve $x^2 = 4ay$, or $y = x^2/4a$ (a *parabola*) on the same axes, and find the area enclosed between the two graphs. (The value of x to be used as integration limit is found by solving the equations of the two curves simultaneously for x, eliminating y.)

12. A square is formed by the coordinate axes and the two lines parallel to the axes through the point $(1, 1)$. Find the ratio of the larger to the smaller of the two areas into which the square is divided by each of the following curves, each of which passes through the point $(1, 1)$ and the origin O:

(a) $y = x$. (b) $y = x^2$. (c) $y = x^3$. (d) $y = x^4$.

B. Lengths of Curves

72. Illustrative Problems. 1. Find by integration the circumference of a circle of radius a.

Solution 1.   This is the curve of Figs. 27 and 35a and the equation gives

$$y = \sqrt{a^2 - x^2}. \tag{a}$$

The length of one quadrant is $\int_0^a \sqrt{1 + \left(\dfrac{dy}{dx}\right)^2}\,dx$ and the circumference is, therefore,

$$C = 4 \int_0^a \sqrt{1 + \left(\frac{dy}{dx}\right)^2}\,dx. \tag{b}$$

From equation (a),

$$\frac{dy}{dx} = -\frac{x}{\sqrt{a^2 - x^2}}.$$

Hence,

$$1 + \left(\frac{dy}{dx}\right)^2 = 1 + \frac{x^2}{a^2 - x^2} = \frac{a^2}{a^2 - x^2},$$

and, therefore, by (b),

$$C = 4a \int_0^a \frac{dx}{\sqrt{a^2 - x^2}}.$$

By integral formula (XXIII), $\displaystyle\int \frac{dx}{\sqrt{a^2 - x^2}} = \sin^{-1}\left(\frac{x}{a}\right)$. Hence

$$C = 4a\left[\sin^{-1}\left(\frac{x}{a}\right)\right]_0^a = 4a\left[\sin^{-1}\left(\frac{a}{a}\right) - \sin^{-1}\left(\frac{0}{a}\right)\right]$$

$$= 4a(\sin^{-1} 1 - \sin^{-1} 0).$$

However, $\sin^{-1} 1 = 90° = \frac{1}{2}\pi$ radians and $\sin^{-1} 0 = 0$. Therefore,

$$C = 4a(\tfrac{1}{2}\pi) = 2\pi a. \tag{a}$$

which is the familiar formula for the circumference of a circle of radius a.

Solution 2.   If a radius is drawn to any point P on the curve in Fig. 27 making a small angle AOP $= \theta$, then from equation (29), article 23, $ds = a\,d\theta$. The length of the quadrant APB is then

$$s = \int_{\theta=0}^{\theta=\angle AOB} ds = \int_{\theta=0}^{\theta=90°} a\,d\theta.$$

But a is constant and $90° = \dfrac{\pi}{2}$. Therefore

$$C = 4a\int_0^{\pi/2} d\theta = 4a[\theta]_0^{\pi/2} = 4a\left[\frac{\pi}{2} - 0\right].$$

$$\therefore \quad C = 2\pi a \tag{b}$$

which is the same as the result (a).

2. Find the formula for the length of the graph of the equation $ay^2 = x^3$ from the origin to any point on the curve.

Solution.   From the equation, $y = \sqrt{\dfrac{x^3}{a}} = \dfrac{1}{\sqrt{a}}x^{3/2}$.

$$\therefore \quad \frac{dy}{dx} = \frac{3}{2\sqrt{a}}x^{1/2} = \frac{3}{2}\sqrt{\frac{x}{a}}$$

$$\sqrt{1 + \left(\frac{dy}{dx}\right)^2} = \sqrt{1 + \frac{9x}{4a}} = \frac{3}{2\sqrt{a}}\sqrt{\frac{4a}{9} + x}$$

$$= \frac{3}{2\sqrt{a}}\left(x + \frac{4a}{9}\right)^{1/2}. \tag{a}$$

Any point P on the curve is designated by the coordinates x, y and at the origin x, $y = 0$. Therefore, using the result (a) in the length integral formula,

$$s = \int_0^x \frac{3}{2\sqrt{a}} \left(x + \frac{4a}{9} \right)^{1/2} dx.$$

$$\therefore \quad s = \frac{3}{2\sqrt{a}} \int_0^x \left(x + \frac{4a}{9} \right)^{1/2} dx.$$

Now, $d\left(x + \frac{4a}{9} \right) = dx$ and the length formula becomes

$$s = \frac{3}{2\sqrt{a}} \int_0^x \left(x + \frac{4a}{9} \right)^{1/2} \cdot d\left(x + \frac{4a}{9} \right)$$

in which the integral is of the form $\int u^{1/2} \, du$ with $u = \left(x + \frac{4a}{9} \right).$
Hence,

$$\int \left(x + \frac{4a}{9} \right)^{1/2} \cdot d\left(x + \frac{4a}{9} \right) = \frac{\left(x + \frac{4a}{9} \right)^{3/2}}{\left(\frac{3}{2} \right)}.$$

$$\therefore \quad s = \frac{3}{2\sqrt{a}} \left[\frac{\left(x + \frac{4a}{9} \right)^{3/2}}{\left(\frac{3}{2} \right)} \right]_0^x = \frac{1}{\sqrt{a}} \left[\left(x + \frac{4a}{9} \right)^{3/2} \right]_0^x$$

$$= \frac{1}{\sqrt{a}} \left[\left(x + \frac{4a}{9} \right)^{3/2} - \left(\frac{4a}{9} \right)^{3/2} \right].$$

Since the $\frac{3}{2}$ power is the same as the square root of the cube this
becomes

$$s = \frac{1}{\sqrt{a}} \left[\sqrt{\left(x + \frac{4a}{9} \right)^3} - \sqrt{\left(\frac{4a}{9} \right)^3} \right].$$

Reducing and simplifying this expression we have finally for the
length of the section OP of the curve (origin to any point P)

$$s = \frac{1}{27} \sqrt{\frac{(9x + 4a)^3}{a}} - \frac{8a}{27}.$$

This curve is called the *semi-cubical parabola* and is historically interesting as the first curve beside the circle whose length was determined.
This was accomplished by William Neil in 1660 without the aid of
the calculus.

3. A certain curve is such that when θ is an angle which varies from
0 to 360°, the coordinates of a point on the curve are given by

$$x = a(\theta - \sin \theta) \tag{a}$$

$$y = a(1 - \cos \theta). \tag{b}$$

Find the length of the curve.

Solution. The length is given by the formula

$$s = \int \sqrt{1 + \left(\frac{dy}{dx}\right)^2} \, dx \tag{c}$$

and in order to find s we must find dy and dx from equations (a) and (b) and substitute in the derivative dy/dx in equation (c).

From (b),

$$dy = a \cdot d(1 - \cos \theta) = a \cdot [-d(\cos \theta)].$$

$$\therefore \quad dy = a \sin \theta \, d\theta, \tag{d}$$

and from (a),

$$dx = a \cdot d(\theta - \sin \theta) = a[d\theta - d(\sin \theta)]$$

$$= a(d\theta - \cos \theta \, d\theta).$$

$$\therefore \quad dx = a(1 - \cos \theta) \, d\theta. \tag{e}$$

Dividing (d) by (e),

$$\frac{dy}{dx} = \frac{\sin \theta}{1 - \cos \theta}. \tag{f}$$

If now we substitute (e) and (f) in the length formula (c),

$$s = \int \sqrt{1 + \left(\frac{\sin \theta}{1 - \cos \theta}\right)^2} \cdot a(1 - \cos \theta) \, d\theta \tag{g}$$

and this expression is to be reduced and integrated. Then,

$$\sqrt{1 + \left(\frac{\sin \theta}{1 - \cos \theta}\right)^2} = \sqrt{1 + \frac{\sin^2 \theta}{(1 - \cos \theta)^2}} = \sqrt{\frac{(1 - \cos \theta)^2 + \sin^2 \theta}{(1 - \cos \theta)^2}}$$

$$= \sqrt{\frac{1 - 2 \cos \theta + \cos^2 \theta + \sin^2 \theta}{(1 - \cos \theta)^2}}.$$

In this, $\cos^2 \theta + \sin^2 \theta = 1$, which allows us to take out 2 as a factor in the numerator. Doing this, and taking the square root, the expression becomes $\sqrt{2} \dfrac{\sqrt{1 - \cos \theta}}{1 - \cos \theta}$. Substituting this in (g), the factor $(1 - \cos \theta)$ cancels and we have

$$s = \int a\sqrt{2}\sqrt{1 - \cos \theta} \, d\theta.$$

Since θ varies from zero to 360° we have finally

$$s = a\sqrt{2} \int_0^{360°} \sqrt{1 - \cos \theta}\ d\theta. \tag{h}$$

In order to integrate $\int \sqrt{1 - \cos \theta}\ d\theta$ we use the following relation from trigonometry: $1 - \cos \theta = 2 \sin^2 (\tfrac{1}{2}\theta)$. Hence $\sqrt{1 - \cos \theta} = \sqrt{2} \sin (\tfrac{1}{2}\theta)$ and formula (h) becomes

$$s = 2a \int_0^{360°} \sin (\tfrac{1}{2}\theta)\ d\theta.$$

Now, $d(\tfrac{1}{2}\theta) = \tfrac{1}{2}\ d\theta$ or $d\theta = 2d(\tfrac{1}{2}\theta)$. Therefore

$$s = 4a \int_0^{360°} \sin (\tfrac{1}{2}\theta)\ d(\tfrac{1}{2}\theta).$$

Integrating,

$$s = -4a \left[\cos \frac{\theta}{2} \right]_0^{360°} = -4a \left[\cos \frac{360}{2} - \cos \frac{0}{2} \right]$$

$$= -4a(\cos 180° - \cos 0°) = -4a(-1 - 1)$$

$$= -4a(-2)$$

$$\therefore \quad s = 8a.$$

This curve is snown in Fig. 36 and is called the *cycloid*. It is the path of a point on the circumference of a circle of radius a which rolls along a straight line. If the circle rolls round the outside of a larger

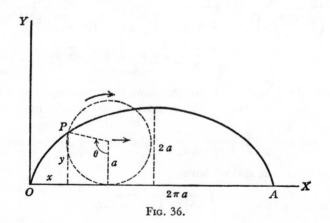

Fig. 36.

circle the path of the point is called an *epicycloid*, and if it rolls round the inside of the larger circle the path is a *hypocycloid*. These three related curves are used in designing gear teeth for the least wear and greatest strength and also in designing the pendulum suspension for very accurate clocks.

4. Find the length of the path of the sliding weight in Problem 6, article 70, from its original position to any point on the path.

Solution. In Fig. 37, as in Fig. 32, OB = PT = a and AP = y.

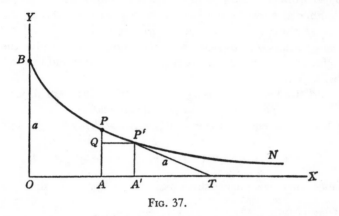

FIG. 37.

Also PP' = ds is a differential of length of the path, PQ = dy, and if PP' is very small PQP' is a right triangle and in the two similar right triangles PQP' and PAT, $\overline{PQ}:\overline{PP'}::\overline{AP}:\overline{PT}$, or

$$\frac{-dy}{ds} = \frac{y}{a}$$

$$\therefore \quad ds = -a\frac{dy}{y}, \tag{a}$$

dy being negative because y decreases from AP to A'P' as x increases from OA to OA'.

At the beginning $y = OB = a$, and the length is to be measured from B to the point P where AP = y. Hence, the length is

$$s = \int_a^y ds = -\int_y^a ds,$$

and, therefore, using (a),

$$s = a \int_y^a \frac{dy}{y} = a[\log_e y]_y^a = a(\log_e a - \log_e y).$$

$$\therefore \quad s = a \log_e \left(\frac{a}{y}\right).$$

This curve is representative of cases in which the expression for $y \, dx$ in the area formula and the expression for ds in the length formula are found from the curve by geometrical methods without having to find y or dy/dx from the equation of the curve. This is due to the fact that the manner of constructing the curve is described. From this same description the equation of the curve could be found and then the values of y and dy/dx used in the usual manner would give the same results for length and area as we have obtained here and in Problem 6, article 70.

5. Find the length of the *logarithmic curve* of Problem 8, article 71, above the OX axis to any point whose abscissa is x.

Solution. The equation is $y = \log_e x$, hence $dy/dx = 1/x$, and

$$s = \int \sqrt{1 + \left(\frac{1}{x}\right)^2} \, dx = \int \frac{1}{x} \sqrt{x^2 + 1} \, dx,$$

and since the curve crosses the x axis at $x = 1$ we have finally

$$s = \int_1^x \frac{1}{x} \sqrt{x^2 + 1} \, dx.$$

Then, by parts,

$$\int \frac{1}{x} \sqrt{x^2 + 1} \, dx = \sqrt{x^2 + 1} - \log_e \left(\frac{1 + \sqrt{x^2 + 1}}{x}\right);$$

therefore,

$$s = \left[\sqrt{x^2 + 1} - \log_e \left(\frac{1 + \sqrt{x^2 + 1}}{x}\right) \right]_1^x.$$

Substituting the limits, subtracting and simplifying, we get finally,

$$s = (\sqrt{x^2 + 1} - \sqrt{2}) - \log_e \left[\frac{1 + \sqrt{x^2 + 1}}{(1 + \sqrt{2})x}\right],$$

which is the formula for the length of the logarithmic curve above the x axis to any point whose abscissa is x.

Note. We add here a problem which, while not an application of the length integral, is still a graphical application of integration.

In article 35 we learned that, if the equation of a curve is given, **we** can find its slope formula by differentiating the equation. Now, since integration is in one sense the reverse of differentiation, we can reverse this process and find the equation which gives the whole curve if we know a formula giving its slope at one point.

The next problem is an illustration of this process and also an example of the meaning of the constant of integration.

6. The slope of a certain curve at a certain point is the reciprocal of twice the ordinate at that point, and the curve passes through the point whose coordinates are (4, 3). Find the equation of the curve.

Solution.   The ordinate of any point is y and the slope at that point is dy/dx. But we are given the slope as $\dfrac{1}{2y}$; hence,

$$\frac{dy}{dx} = \frac{1}{2y} \quad \text{and} \quad 2y\, dy = dx.$$

Integrating,

$$2 \int y\, dy = \int dx$$

$$\therefore \quad y^2 = x + C. \tag{a}$$

This is a relation between the coordinates x, y of any point on the curve and is, therefore, the equation of the curve, but since C is unknown we do not know the full equation. This means that (a) may represent any one of a number of curves, each of which corresponds to a particular value of the constant C. We know, however, that the curve whose equation we wish to find passes through the point at which $x = 4$, $y = 3$, and therefore these values of x, y must satisfy the equation (a) for that one curve through this point.

Substitution of these values in the equation (a) gives

$$9 = 5 + C.$$

$$\therefore \quad C = 4,$$

and this value of C in (a) gives finally the complete equation of the particular curve,

$$y^2 = x + 4.$$

Thus by specifying a condition which the integral (a) must satisfy the value of the integration constant can be found. This is the case in general.

73. Problems for Solution.

Find the length of the graph of each of the following equations between the two points whose abscissas are given in parentheses:

1. $y = \frac{2}{3}(4 - x)^{3/2}$ $(x = 0, 4)$.
2. $y^2 = 4x - x^2$ $(x = 0, 1)$.
3. $y^2 = x^3$ $(x = 0, \frac{5}{9})$.
4. $y = \log_e (\sec x)$ $\left(x = 0, \dfrac{\pi}{3}\right)$.

5. Show by integration that the square of the hypotenuse of a right triangle equals the sum of the squares of the legs. (Suggestion: Use Fig. 31.)

6. Derive the integral formula for the length of the bend of the river in Problem 7, article 71, and calculate the length.

7. The slope of the tangent to a certain curve at any point, x, y is $-(4/9)(x/y)$ and the curve passes through the point $(3,2)$. Show that its equation is $x^2/18 + y^2/8 = 1$ and by comparison with Problem 9, article 71, determine the nature of the curve.

8. By using the result of Problem 9, article 71, or by direct integration, find the area of the figure of Problem 7 above.

9. In Problem 3 of article 72, solve equation (b) for $1 - \cos \theta$, and also for $\cos \theta$, and from this value of $\cos \theta$ find $\sin \theta$ by trigonometric formula. Substitute these values in formula (f) and so express dy/dx in terms of y. Then use the length integral formula $s = 2 \displaystyle\int_0^{2a} \sqrt{1 + (dx/dy)^2}\, dy$ to find the length of one arch of the *cycloid* of that problem.

10. The form of the curve of a chain or telephone wire hanging between two points of support (poles) is given by the equation $y = \frac{1}{2}a \left(e^{\frac{x}{a}} + e^{-\frac{x}{a}} \right)$, where a is the height of the lowest point above the x axis. The length of wire or chain from the lowest point to any point which is at a horizontal distance x from the lowest point is given by the formula $\displaystyle\int_0^x ds$. Calculate this length by integration. (This curve is called the *catenary*.)

Chapter 17

USE OF INTEGRALS
IN SOLVING PROBLEMS

74. Introduction. The preceding chapter contains illustrations of the use of integrals in connection with graphs, having to do mainly with areas and lengths of curves. In many technical and scientific problems involving graphs, the area, the length or the slope of a curve is of great importance. Even though we may not have to do with graphs directly, however, the interpretation of an integral as a sum is of the greatest importance, as has been stated in article 68. In the following articles are given a number of problems which involve definite integrals and illustrate the truth of this statement. There are others which make use of integration simply as the reverse of differentiation and involve the constant of the indefinite integral.

In these problems, use will be made of the formulas of article 56 and also of several formulas derived in the solutions of examples and problems in the preceding articles. For convenience of reference, these latter are collected and listed here.

(XXII) $$\int \frac{dx}{x^2 + a^2} = \frac{1}{a} \tan^{-1} \left(\frac{x}{a} \right).$$

(XXIII) $$\int \frac{dx}{\sqrt{a^2 - x^2}} = \sin^{-1} \left(\frac{x}{a} \right).$$

(XXIV) $$\int \frac{dx}{\sqrt{x^2 \pm a^2}} = \log_e (x + \sqrt{x^2 \pm a^2}).$$

(XXV) $$\int \sqrt{x^2 \pm a^2} \, dx = \tfrac{1}{2}[x\sqrt{x^2 \pm a^2} \pm a^2 \log_e (x + \sqrt{x^2 \pm a^2})].$$

(XXVI) $$\int \sqrt{a^2 - x^2} \, dx = \frac{1}{2} \left[x\sqrt{a^2 - x^2} + a^2 \sin^{-1} \left(\frac{x}{a} \right) \right].$$

75. Population Increase. *The normal increase of population at any time is proportional to the population at that time. If the population of a*

183

country at a certain time and its increase in a certain period are known, derive a formula by means of which the population can be calculated in advance for any time thereafter (barring immigration and national calamities).

Solution. Let P represent the population at any time t; then the rate of increase at that time is dP/dt. For one quantity to be proportional to a second quantity it must be equal to the second multiplied by a constant. If, therefore, dP/dt is proportional to P,

$$\frac{dP}{dt} = kP \tag{a}$$

where k is the proportionality constant. Transforming (a),

$$\frac{dP}{P} = k\,dt.$$

Integrating,

$$\int \frac{dP}{P} = k \int dt.$$

$$\therefore \quad \log_e P = kt + C.$$

In order to obtain the final computing formula the value of C must be determined. Now at the time which we take as the starting time the elapsed time is zero, that is, $t = 0$, and at that time the population is known: let it be P_0. These two values of P and t in (b) give

$$\log_e P_0 = 0 + C.$$

$$\therefore \quad C = \log_e P_0,$$

and this value of C in (b) gives

$$\log_e P = kt + \log_e P_0.$$

Transposing,

$$\log_e P - \log_e P_0 = kt$$

or,

$$\log_e (P/P_0) = kt.$$

Comparing this result with formulas (53), article 44, we see that

$$(P/P_0) = e^{kt}.$$

$$\therefore \quad P = P_0 e^{kt}. \tag{c}$$

This is the desired formula. In order to use it, however, the value of k must be known.

In order to determine k consider the population at the time when the

increase in a certain period is known, as 100% or 1, and use this value
(1) in equation (a).   With $P = 1$ this gives $dP/dt = k$, or,

$$k = \left(\frac{dP}{dt}\right)_{P=1}.$$

This means that k is the rate of increase when P is 1 or 100%, that is,
k is the *fraction* or percentage of P added in the unit period of time.   If
the time unit is the year, then k is the fractional or percentage increase
per year expressed as the fraction or percentage of the population exist-
ing at the beginning of the year.   According to the statement of the
given problem, therefore, P_0 and k are known quantities.

By means of formula (c), therefore, the population P can be calculated at any time t years from any chosen starting time.   If immigration, war, fatal epidemics, etc., are considered, the formula is modified
but can be obtained in the same general way.

Such formulas as (c) are used in connection with the census and for
taxation and parliamentary representation, etc.   The same kind of formula also holds good in many other natural processes of growth and
decay.   (See the following chapter and also certain other problems in
this chapter.)

76. The Laws of Falling Bodies. *An object is allowed to fall freely
toward the earth.   It is required to find its velocity and its distance from the
starting point at any instant thereafter when the effect of air resistance is disregarded.*

Solution.   When the object or body is released without being thrown,
no force acts on it except that of gravitation, and as this force acts
constantly, the body is steadily accelerated, so that in each second its
velocity increases 32.16 feet per second.   This acceleration is called
the *acceleration of gravity* and denoted by the letter g.   Hence $g = 32.16$
(ft/sec.) per second.

Now we have found in Chapter 6 that when s is space or distance
and t is time, then $\frac{d^2s}{dt^2}$ is acceleration.   Hence, in the present case we
have $\frac{d^2s}{dt^2} = g$, and g is a constant.   This can be written as

$$\frac{d}{dt}\left(\frac{ds}{dt}\right) = g.$$

$$\therefore \quad d\left(\frac{ds}{dt}\right) = g\,dt.$$

Integrating,

$$\int d\left(\frac{ds}{dt}\right) = g \int dt.$$

$$\therefore \quad \frac{ds}{dt} = gt + C. \tag{a}$$

Now ds/dt is velocity, denoted by v. Therefore

$$v = gt + C. \tag{b}$$

In order to determine C we again, as in the preceding problem, make use of the starting or *initial* conditions. At the instant when the object is released no time has elapsed and it has no velocity. That is, $t = 0$ and $v = 0$. Using these values in (b) we find $C = 0$ and therefore

$$v = gt \tag{c}$$

is the velocity in feet per second at any time t seconds after the instant of release.

In order to find the distance s through which the body has fallen at any time t, consider equation (a). Since we found that C is zero this equation becomes $ds/dt = gt$. Hence,

$$ds = gt\, dt.$$

Integrating.

$$\int ds = g \int t\, dt.$$

$$\therefore \quad s = \tfrac{1}{2}gt^2 + C', \tag{d}$$

and C' is to be determined.

Again making use of the initial conditions when t is zero so also is the distance s. Putting $t = 0$, $s = 0$ in (d) we find that C' is zero and therefore (d) becomes

$$s = \tfrac{1}{2}gt^2. \tag{e}$$

From this result we find by solving for t that

$$t = \sqrt{\frac{2s}{g}} = \frac{1}{4}\sqrt{s}$$

is the time in seconds required for the body to fall through a height s in feet. Thus (disregarding air resistance, which is small for small objects at moderate velocities) if a stone is dropped from the top of the

Washington Monument, which is 550 feet high, $s = 550$ and the time required for it to reach the ground is

$$t = \tfrac{1}{4}\sqrt{550} = 5.86 \text{ sec.}$$

According to formula (c), it will, when it reaches the ground, have a velocity of about

$$v = 32.2 \times 5.86 = 189 \text{ ft./sec.,}$$

or, since 88 ft. per sec. = 1 mile per min.,

$$v = \tfrac{189}{88} = 2.16 \text{ mi./min.}$$

Suppose, now, that instead of being allowed to fall freely, the object is thrown or shot straight downward with a velocity V. After it starts gravity alone acts on it and we again have equation (b). Now however, when $t = 0$, $v = V$ and (b) gives $C = V$, and instead of (c) we get from (b),

$$v = gt + V. \tag{f}$$

Equation (a) in this case becomes

$$\frac{ds}{dt} = gt + V.$$

$$\therefore \quad ds = gt\,dt + V\,dt.$$

Integrating,

$$\int ds = g \int t\,dt + V \int dt.$$

$$\therefore \quad s = \tfrac{1}{2}gt^2 + Vt + C''.$$

We still have however, when $t = 0$, $s = 0$, and these values in the last equation give $C'' = 0$ and therefore

$$s = \tfrac{1}{2}gt^2 + Vt.$$

Since $g = 32.16$, $\tfrac{1}{2}g = 16.08$ or about 16 and the last formula becomes

$$s = 16t^2 + Vt. \tag{g}$$

According to this formula, if the stone were thrown downward from the top of the Washington Monument at a velocity of, say, 20 feet per second, we have $s = 550$, $V = 20$ and hence

$$550 = 16t^2 + 20t,$$

or,

$$8t^2 + 10t - 275 = 0$$

and in order to find the time required for the stone to reach the ground this equation must be solved for t.

This is a quadratic equation and according to algebra the solution is

$$t = \frac{-10 \pm \sqrt{(10)^2 - 4 \times 8(-275)}}{2 \times 8} = \frac{-10 \pm \sqrt{8900}}{16}$$

$$= \frac{10}{16}(-1 \pm \sqrt{89}) = \frac{5}{8}(-1 \pm 9.434) = 5.27; \; -6.52;$$

and since the negative value has no physical meaning, the time required is about

$$t = 5.3 \text{ sec.}$$

as compared with 5.9 seconds for the free fall. By throwing it with a greater initial velocity, that is, making V greater in equation (g), the time would be still less.

When thrown with the initial velocity $V = 20$ and falling for a time $t = 5.3$ equation (f) gives for the final velocity

$$v = (32.2 \times 5.3) + 20 = 191 \text{ ft./sec.}$$

as compared with 189 for the free fall.

By application of the principles and methods of this problem information is obtained concerning bodies falling to the earth from outer space (meteors).

77. Path and Range of Projectiles. *A body is projected from the earth's surface with a known velocity and direction. It is required to find its path, time of flight, height reached, and distance travelled.*

Solution. Let the initial velocity be V feet per second and let the initial direction make an angle α with the earth's surface, supposed horizontal and plane, and disregard the effect of air resistance. The projectile will rise for a time, reach a height H, and then begin to fall, and after a certain "time of flight" (T) strike the earth at a distance R (the "range") from the starting point. The path will then be something like that shown in Fig. 38, and in order to find the path, height, and range we must derive the equation of the curve OPA and calculate the distances PF and OA.

If there were no forces acting on the projectile after it leaves the point O, it would go on in the straight line OB forever, but it is acted on by gravity and like the falling bodies already discussed it falls steadily so that it passes through P instead of B, having in effect fallen through the distance BP by the time it reaches P.

Now the original velocity V can be considered as made up of two component velocities, v_x in the horizontal or OX direction and v_y in the vertically upward direction OY.   If there were a force acting on

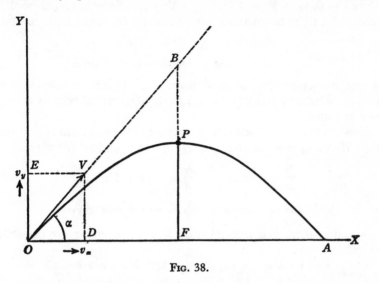

Fig. 38.

the object in the horizontal direction there would be an acceleration in this direction, but as there is no horizontal force the horizontal acceleration, dv_x/dt, is zero.   That is,

$$\frac{dv_x}{dt} = 0. \tag{a}$$

There is, however, a vertical acceleration.   It is the acceleration of gravity g and it causes the falling discussed above.   But this acceleration is downward while the original vertical velocity is upward.   Therefore, the vertical acceleration, since it tends to decrease v_y, is negative and equal to g.   Hence,

$$\frac{dv_y}{dt} = -g. \tag{b}$$

From these two equations (a), (b), we get

$$dv_x = 0, \quad dv_y = -g\, dt.$$

Integrating,

$$v_x = C, \quad v_y = -gt + C'. \tag{c}$$

Now from the figure, at the instant of projection when the total velocity is V in the direction shown, the x-component is $OD = V \cos \alpha$, and the y-component is $OE = V \sin \alpha$. That is, at the initial moment, when $t = 0$, $v_x = V \cos \alpha$ and $v_y = V \sin \alpha$. Substituting these three values of t, v_x, v_y in equations (c) we find for the integration constants

$$C = V \cos \alpha, \quad C' = V \sin \alpha.$$

$$\therefore \quad v_x = V \cos \alpha, \quad v_y = -gt + V \sin \alpha \tag{d}$$

at any time t after the instant of projection. It is to be noted that the horizontal velocity remains constant while the vertical velocity changes with the time.

Now velocity is the time rate of distance in the direction of the velocity. Hence $v_x = dx/dt$ and $v_y = dy/dt$, and equations (d) become

$$\frac{dx}{dt} = V \cos \alpha, \quad \frac{dy}{dt} = -gt + V \sin \alpha;$$

$$\therefore \quad dx = V \cos \alpha \, dt, \quad dy = -gt \, dt + V \sin \alpha \, dt.$$

Integrating these and remembering that V, α, and g are constants, we get

$$x = V \cos \alpha \cdot t + C_1, \quad y = -\tfrac{1}{2}gt^2 + V \sin \alpha \cdot t + C_2.$$

To find C_1 and C_2 we make use of the conditions that when $t = 0$ both x and y are zero (the projectile being then at O). These values substituted in the last equations give us zero for both the integration constants and therefore the equations become

$$x = V \cos \alpha \cdot t, \quad y = -\tfrac{1}{2}gt^2 + V \sin \alpha \cdot t. \tag{e}$$

These give the horizontal distance x and the vertical height y of the projectile at any time t after the instant of projection.

In order to find the equation of the path we must find an equation giving the relation between x and y which shall apply for all values of x and y. To do this we must combine the two equations (e) into one equation containing x and y and the constants V and α but not containing t. That is, as we say in algebra, we must eliminate t from the equations (e). To do this we solve the first for t getting

$$t = \frac{x}{V \cos \alpha}, \tag{f}$$

and substitute this value in the second equation (e), obtaining

$$y = -\frac{g}{2}\left(\frac{x}{V\cos\alpha}\right)^2 + V\sin\alpha\left(\frac{x}{V\cos\alpha}\right)$$

$$= -\frac{g}{2}\frac{x^2}{V^2\cos^2\alpha} + \frac{\sin\alpha\cdot x}{\cos\alpha}$$

$$= -\frac{gx^2}{2V^2\cos^2\alpha} + \tan\alpha\cdot x.$$

$$\therefore\quad y = (\tan\alpha)x - \left(\frac{g}{2V^2\cos^2\alpha}\right)x^2. \tag{g}$$

This is the equation of the path OPA of the projectile, Fig. 38. This curve is called a *parabola* and is of the general shape shown. If V and α are changed the height PF and distance OA change also so that the curve becomes flatter or steeper, but it retains the same characteristic form.

In order to find the range, OA, we notice that at O and at A, when the projectile is at the surface, the value of y is zero. If, therefore, we put $y = 0$, in equation (g), we shall have an equation containing only the unknown x and when $y = 0$ at O, x is also zero, but when $y = 0$ at A then $x = R$, where OA $= R$ is the range. When $y = 0$ equation (g) gives,

$$0 = (\tan\alpha)x - \left(\frac{g}{2V^2\cos^2\alpha}\right)x^2.$$

$$\therefore\quad x\left[\tan\alpha - \left(\frac{g}{2V^2\cos^2\alpha}\right)x\right] = 0$$

and either $x = 0$ or $\tan\alpha - \left(\frac{g}{2V^2\cos^2\alpha}\right)x = 0$.

As just shown, the first solution applies at O, the second at A. From the second we get,

$$\left(\frac{g}{2V^2\cos^2\alpha}\right)x = \tan\alpha$$

and

$$x = \frac{2V^2}{g}\cos^2\alpha\cdot\tan\alpha = \frac{2V^2}{g}\cos^2\alpha\frac{\sin\alpha}{\cos\alpha}.$$

$$\therefore\quad x = \frac{2V^2}{g}\sin\alpha\cos\alpha \tag{h}$$

or,

$$R = \frac{2V^2}{g}\sin\alpha\cos\alpha. \tag{i}$$

Since 2 sin α cos α = sin (2α), this last equation can be written,

$$R = \frac{V^2}{g} \sin 2\alpha. \tag{i'}$$

Either (i) or (i') gives the range for any initial velocity V and angle of projection α. In gunnery V is called the "muzzle velocity" and α the "elevation."

To find the time of flight, the time required for the projectile to reach A, we put the value of x given by (h), in equation (f). Then,

$$t = \frac{2V^2}{g} \frac{\sin \alpha \cos \alpha}{V \cos \alpha}.$$

$$\therefore \quad T = \frac{2V}{g} \sin \alpha. \tag{j}$$

So far, V and α are constant. Suppose, now, that with a fixed muzzle velocity V (determined by the gun and powder charge) it is desired to find the elevation which gives a maximum range, that is, the best value of α. In this case V is constant and α is variable, and in equation (i') we are to find the value of α which makes R a maximum. According to Chapter 8 this means that $dR/d\alpha = 0$. Differentiating (i'),

$$\frac{dR}{d\alpha} = \frac{2V^2}{g} \cos 2\alpha,$$

and for this to be zero, we must have, since $2V^2/g$ is not zero,

$$\cos 2\alpha = 0, \quad \therefore \quad 2\alpha = 90°$$

and

$$\alpha = 45°$$

is the elevation which gives the maximum range. Using this value in (i') we get for the maximum range,

$$R_m = \frac{V^2}{g}, \tag{k}$$

and since g is constant the maximum range is proportional to the square of the muzzle velocity. At an elevation of 45°, therefore, if the muzzle velocity should be doubled the maximum range would theoretically be multiplied four times. Of course the greatest muzzle velocities attainable, consistent with the safety of the gun, are used.

To find the height to which the projectile rises, we have in Fig. 38,

$\overline{OF} = \frac{1}{2}\overline{OA}$ and, therefore, at F the abscissa is half that given by equation (h).   Therefore, at F, $x = (V^2/g)\sin\alpha\cos\alpha$.   Substituting this value of x in (g) we find that at F,

$$y = \tan\alpha\left(\frac{V^2}{g}\sin\alpha\cos\alpha\right) - \left(\frac{g}{2V^2\cos^2\alpha}\right)\left(\frac{V^2}{g}\sin\alpha\cos\alpha\right)^2$$

$$= \frac{V^2}{g}\sin\alpha\cos\alpha\,\frac{\sin\alpha}{\cos\alpha} - \frac{V^2}{2g}\frac{\sin^2\alpha\cos^2\alpha}{\cos^2\alpha}.$$

$$\therefore \quad \overline{PF} = \frac{V^2}{g}\sin^2\alpha - \frac{V^2}{2g}\sin^2\alpha,$$

or,

$$H = \frac{V^2}{2g}\sin^2\alpha.$$

When $\alpha = 45°$, $\sin^2\alpha = (1/\sqrt{2})^2 = \frac{1}{2}$ and the height corresponding to the maximum range is from the last formula,

$$H_m = \frac{V^2}{4g}.$$

Comparing this with the formula (k) for the maximum range, it is seen that,

$$H_m = \frac{1}{4}R_m. \tag{l}$$

In the preceding pages we have not considered the resistance of the air nor the curvature and rotation of the earth.   In actual gunnery, however, these must be considered and require modifications to be made in the formulas.

Supposing, however, for the sake of practice in using the formulas, that the earth is approximately flat for a distance of 72 miles and that in very long range firing the projectile goes to such a height that over most of its path the air resistance is small, it is interesting to apply our results to the case of the long range gun which bombarded Paris during the First World War at a range of 72 miles.

It may be assumed that 72 miles was the maximum range or the gun would have been even farther away.   Then $R_m = 72$ miles and by formula (l) the shell rose to a hieght of about

$$H_m = \frac{1}{4}(72) = 18 \text{ miles.}$$

Now, 72 miles = 380,000 feet and $g = 32$ (ft./sec.) per sec.   Hence, by formula (k), $380,000 = V^2/32$, or $V^2 = 32 \times 380,000$ and hence

$$V = 3490 \text{ ft./sec.} = .661 \text{ mi./sec.}$$

was the muzzle velocity of the shell.

Assuming $\alpha = 45°$ when the range is a maximum, formula (j) gives for the time of flight

$$T = (\sqrt{2}/g)V = .044V$$

and in the present case

$$T = .044 \times 3490 = 154 \text{ sec.} = 2\tfrac{1}{2} \text{ min.}$$

Of course the actual values of height, muzzle velocity and time of flight were greater than the values we have obtained on account of the air resistance to be overcome.

This article is long but the subject is of such interest and the problem gives applications of the calculus which are of such beauty that it has been discussed at some length.   In actual gunnery, or *ballistics*, however, many other considerations enter in addition to those entering here and the so-called "parabolic theory" is only used as a theoretical basis.

78. Length of Belting or Paper in a Roll.   *A spiral line begins at a distance a from a center point and ends at a distance b, the distance between turns being c.   Find the total length of the line.*

Solution.   In Fig. 39, O is the center, A is the beginning of the spiral,

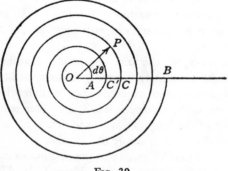

FIG. 39.

B the end, OA $= a$, OB $= b$ and AC$' = c$.   If we think of any particular turn at a distance OC$'$ from the center, then in going from A once around and to C$'$ the angle BOP $= \theta$ increases through $360° = 2\pi$ radians, and the radial increase along OB is c.   In turning through any small angle $d\theta$ therefore the radial increase is dr (where r is the radial distance), and we have the proportion $d\theta : 2\pi :: dr : c$, or

$$\frac{d\theta}{dr} = \frac{2\pi}{c}. \tag{a}$$

Consider, now, any layer or turn at any distance $OC = r$ from the center and in order to obtain a formula for the differential of length $ds = CP$ corresponding to the small angle $d\theta$, Fig. 39, let us exaggerate somewhat the inclination of the layer or turn to the line OB, as in Fig. 40.   Draw the arc CQ with O as center and OC as radius.   Then

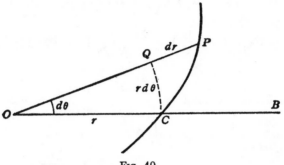

Fig. 40.

$OQ = OC = r$ and $QP = dr$, the increase in r due to the radial line OP turning through the small angle $d\theta$, and according to equation (29), article 23, the arc $CQ = r\,d\theta$.

If now $d\theta$ is very small, so also are $r\,d\theta$, dr and ds, and since the angle CQP is a right angle, we may consider CQ and CP as straight and the triangle CQP as a right triangle.   Then $\overline{CP}^2 = \overline{PQ}^2 + \overline{QC}^2$, or

$$(ds)^2 = (dr)^2 + (r\,d\theta)^2.$$

Dividing this by $(dr)^2$,

$$\left(\frac{ds}{dr}\right)^2 = 1 + r^2 \left(\frac{d\theta}{dr}\right)^2.$$

Taking the square root,

$$\frac{ds}{dr} = \sqrt{1 + r^2 \left(\frac{d\theta}{dr}\right)^2}.$$

$$\therefore \quad ds = \sqrt{1 + r^2 \left(\frac{d\theta}{dr}\right)^2} \cdot dr.$$

This gives the differential of length in terms of dr and as the spiral continues, r changes from OA to OB in Fig. 39. The total length is, therefore, $s = \int_{OA}^{OB} ds$. Hence,

$$s = \int_a^b \sqrt{1 + r^2 \left(\frac{d\theta}{dr}\right)^2}\, dr. \tag{b}$$

This is the desired formula for the length but it cannot be integrated until we know $d\theta/dr$. But this is given by equation (a), therefore,

$$s = \int_a^b \sqrt{1 + r^2 \left(\frac{2\pi}{c}\right)^2}\, dr. \tag{c}$$

By algebra $\sqrt{1 + r^2 \left(\frac{2\pi}{c}\right)^2} = \frac{2\pi}{c} \sqrt{\left(\frac{c}{2\pi}\right)^2 + r^2}$ and if we write

$$k = \frac{c}{2\pi}, \tag{d}$$

then equation (c) becomes

$$s = \frac{1}{k} \int_a^b \sqrt{r^2 + k^2}\, dr$$

and $\int \sqrt{r^2 + k^2}\, dr$ is the same form as $\int \sqrt{x^2 + a^2}\, dx$ in formula (XXV). Therefore,

$$s = \frac{1}{2k} [r\sqrt{r^2 + k^2} + k^2 \log_e (r + \sqrt{r^2 + k^2})]_a^b.$$

Substituting the values b and a, subtracting, and simplifying, this becomes finally,

$$s = \frac{1}{2k} \left[(b\sqrt{b^2 + k^2} - a\sqrt{a^2 + k^2}) + k^2 \log_e \left(\frac{b + \sqrt{b^2 + k^2}}{a + \sqrt{a^2 + k^2}}\right) \right]. \tag{e}$$

This formula is somewhat complicated and looks even more complicated than it is. A very closely approximate formula can be obtained from it, however, which is much simpler.

Since by (d) $k = c/2\pi$, then if we are considering leather belting $\frac{1}{8}$ inch thick wound on a wooden spool 4 inches in diameter, $c = \frac{1}{8}$, $a = \frac{1}{2}(4) = 2$, $k = \frac{\frac{1}{8}}{2\pi} = \frac{1}{16\pi} = .0199$ and $k^2 = .000395$, while $a^2 = 4$ and of course b^2, the square of the outer diameter of the roll, is greater.

Therefore, under the square root signs we may disregard k^2 in comparison with a^2 and b^2 in any ordinary spiral roll, and write $\sqrt{a^2 + k^2} = \sqrt{a^2} = a$ and $\sqrt{b^2 + k^2} = b$.

Formula (e) then becomes

$$s = \frac{1}{2k}\left[(b \cdot b - a \cdot a) - k^2 \log_e \left(\frac{b + b}{a + a} \right) \right]$$

or,

$$s = \frac{1}{2k}\left[(b^2 - a^2) - k^2 \log_e \left(\frac{b}{a} \right) \right]. \tag{f}$$

This can again be simplified.   If b is not very large in comparison with a then b/a is a small number and $\log_e (b/a)$ is very small.   Also, as we saw above, if $c = \frac{1}{8}$, $k^2 = .0004$ and if c is very small, as in the case of paper in rolls, then k^2 is almost zero and $k^2 \log_e (b/a)$ may be disregarded.   Formula (f) then reduces to $s = (b^2 - a^2)/2k$, and putting in the value of k from (d), we get finally,

$$s = \frac{\pi}{c} (b^2 - a^2). \tag{g}$$

As an illustration of the use of this formula let us calculate the length of leather belting $\frac{1}{8}$ inch thick which is wound tightly on a spool or roller 4 inches in diameter to make a roll 2 feet in diameter.   In this case $a = 2$, $b = 12$, $c = \frac{1}{8}$ inches and

$$s = \frac{\pi}{\left(\frac{1}{8}\right)} [(12)^2 - (2)^2] = 8\pi(144 - 4) = 8 \times 3.1416 \times 140.$$

∴ $s = 3530$ inches $= 294$ feet.

If paper .005 inch thick is wound on a cardboard tube $1\frac{1}{2}$ inches in diameter to make a roll 5 inches in diameter, we have $a = .75$, $b = 2.5$, $c = .005$ inches, and the calculation gives

$$s = 3574 \text{ inches} = 298 \text{ feet.}$$

79. Charge and Discharge of an Electric Condenser.   *A known voltage is applied to a condenser of known capacity in series with a known resistance.   What is the charge in the condenser at any time after the voltage is applied or withdrawn?*

Solution.   Let E be the applied voltage, C the condenser capacity and R the series resistance, and let q be the charge at any time t after the charge begins.

The total voltage is consumed in forcing current through the resist-

ance and in charging the condenser.   By Ohm's Law the voltage used
up in the resistance is $E_r = iR$ where i is the current, and according
to the principle of the condenser the voltage across its plates at any
instant is equal to the quotient of the charge by the capacity, $E_c = q/C$.
Since the total voltage is consumed in these two parts, $E_r + E_c = E$,
that is,

$$Ri + \frac{q}{C} = E. \tag{a}$$

Now by definition electric current is the rate of flow of charge, that is,
$i = dq/dt$.   Hence (a) becomes

$$R\frac{dq}{dt} + \frac{q}{C} = E, \tag{b}$$

and in this equation R, C, E are constants.

To transform (b) for integration first transpose and divide by R;
this gives

$$\frac{dq}{dt} = \frac{1}{R}\left(E - \frac{q}{C}\right) = -\frac{1}{R}\left(\frac{q}{C} - E\right).$$

$$\therefore \quad \frac{dq}{dt} = -\frac{q - CE}{RC}.$$

Next multiply by dt and divide by $(q - CE)$; we then have

$$\frac{dq}{q - CE} = -\frac{1}{RC}\,dt.$$

Integrating,

$$\int \frac{dq}{q - CE} = -\frac{1}{RC}\int dt.$$

But $d(q - CE) = dq$, so we can write

$$\int \frac{d(q - CE)}{q - CE} = -\frac{1}{RC}\int dt$$

and the first integral is of the form $\int \dfrac{dx}{x} = \log_e x.$ Hence

$$\log_e (q - CE) = -\frac{1}{RC}t + C'. \tag{c}$$

Now when the voltage is applied, that is, at the instant when $t = 0$,
there is not yet any charge in the condenser, $q = 0$.   These values of
t, q in (c) give

$$\log_e (0 - CE) = 0 + C'$$
$$C' = \log_e (-CE)$$

and (c) becomes

$$\log_e (q - CE) = -\frac{t}{RC} + \log_e (-CE)$$

or,

$$\log_e (q - CE) - \log_e (-CE) = -\frac{t}{RC}.$$

$$\therefore \quad \log_e \left(\frac{q - CE}{-CE}\right) = -\frac{t}{RC}$$

$$\therefore \quad \frac{q - CE}{-CE} = e^{-\frac{t}{RC}}$$

and

$$q - CE = -CEe^{-\frac{t}{RC}}.$$

Transposing and factoring this we get finally

$$q = CE \left(1 - e^{-\frac{t}{RC}}\right), \tag{d}$$

which gives the charge q at any time t after charging begins, in terms of C, E, R which are known constants.

If the voltage is removed and the condenser and resistance are connected directly in series, there is then no applied voltage, that is, $E = 0$, and equation (b) becomes

$$R\frac{dq}{dt} + \frac{q}{C} = 0.$$

$$\therefore \quad \frac{dq}{dt} = -\frac{q}{CR} \quad \text{or} \quad dq = -\frac{q}{CR}dt$$

and

$$\frac{dq}{q} = -\frac{1}{CR}dt.$$

Integrating,

$$\log_e q = -\frac{t}{RC} + C''. \tag{e}$$

At the moment when the voltage is removed the condenser is fully charged with the charge Q, that is, when $t = 0$, $q = Q$. These values in (e) give

$$\log_e Q = 0 + C''. \quad \therefore \quad C'' = \log_e Q$$

and

$$\log_e q = -\frac{t}{CR} + \log_e Q,$$

or,

$$\log_e q - \log_e Q = -\frac{t}{CR}.$$

$$\therefore \quad \log_e \left(\frac{q}{Q}\right) = -\frac{t}{CR} \quad \text{and} \quad \frac{q}{Q} = e^{-\frac{t}{CR}}.$$

$$\therefore \quad q = Qe^{-\frac{t}{CR}} \tag{f}$$

which states that at any time t after the discharge begins the charge is equal to the original charge Q multiplied by the $-(t/CR)$ power of the number e.

Let us see what is the charge at a time $t = CR$. Using this value of t in (f) we get $q = Qe^{-1} = Q/e$.

$$\therefore \quad q = .369Q.$$

This means that the charge has in CR seconds dropped to 36.9% of its original value, that is, it has decreased by 63.1%, or nearly two thirds. The quantity CR is called the *time constant* of the condenser resistance circuit.

For a fully charged condenser $Q = CE$, and, therefore, we can write equation (d) as,

$$q = Q\left(1 - e^{-\frac{t}{RC}}\right).$$

If in this we let $t = CR$ we find,

$$q = .631Q,$$

which means that in *charging*, the condenser is about two thirds charged in the time equal to the time constant.

80. Rise and Fall of Electric Current in a Coil. By methods similar to those used in the preceding article we can *find the formula giving the current at any instant after a voltage is applied to an inductive coil and a resistance in series.*

Solution. Let E be the applied voltage, L the inductance of the coil and R the series resistance, and let i be the current at any time t after the voltage is applied. The drop across the resistance is by Ohm's Law equal to Ri and that across the coil is equal to the product of the inductance by the rate of change of the current, $L(di/dt)$; and the sum of these two must equal the total applied voltage. Hence

$$L\frac{di}{dt} + Ri = E. \tag{a}$$

Comparing this with equation (b) for the condenser circuit, namely,

$$R\frac{dq}{dt} + \frac{1}{C}q = E,$$

we see that they are of the same form with L replacing R, R replacing $1/C$ and i replacing q. The two are integrated in the same manner, therefore, and corresponding to (d) of the preceding article we have,

$$i = \frac{E}{R}\left(1 - e^{-\frac{Rt}{L}}\right), \tag{b}$$

which gives the value of the current i in the coil and resistance at any time t after the switch is closed, in terms of the known constants E, L, R.

Similarly, also, when the voltage is removed and the coil and resistance joined in series, the current begins to die away. In this case $E = 0$ and equation (a) becomes

$$L\frac{di}{dt} + Ri = 0,$$

which like the corresponding equation for the condenser gives when integrated

$$i = \frac{E}{R}e^{-\frac{Rt}{L}}. \tag{c}$$

Since in the steady condition, when the full current I is flowing, Ohm's Law gives $E/R = I$, we can write equation (b) for the rising current, and (c) for the falling current as

$$i_r = I\left(1 - e^{-\frac{Rt}{L}}\right), \quad i_f = Ie^{-\frac{Rt}{L}}, \tag{d}$$

and by the method used with the condenser the *time constant* of an inductive circuit is found to be L/R. This means that in a circuit consisting of a coil and a resistance in series the current rises to about two thirds of its final full value in a time L/R seconds after the voltage is applied and falls to about one third that full value in L/R seconds after the coil and resistance are short circuited without the voltage.

81. Differential Equations. Equation (a) above and (b) of article 79 are called *differential equations* because they are equations involving the differentials of the variables q, i and t. When we get rid of the differentials by integration so that only the variables (and constants) and not their differentials appear in the final equations, we are said to have *solved* the differential equation.

The above differential equations are of very great importance in the theory of alternating currents, telephony, power transmission and radio transmission and reception. An even more important differential equation is the one which arises when not only resistance and capacity or resistance and inductance, but resistance, inductance and capacity are all three in the circuit.

In this case, as before, the voltage consumed in the resistance is Ri, that in the coil is $L(di/dt)$, and that in the condenser is q/C, and since the total is E we have

$$L\frac{di}{dt} + Ri + \frac{1}{C}q = E. \tag{a}$$

In order to have only one variable besides the time t, we must eliminate either q or i by expressing it in terms of the other. We eliminate q as follows. Differentiate (a) with respect to t, remembering that L, R, C, E are constants:

$$L\frac{d^2i}{dt^2} + R\frac{di}{dt} + \frac{1}{C}\frac{dq}{dt} = 0.$$

But, as stated above, $dq/dt = i$. Substituting this in the last equation, we get

$$L\frac{d^2i}{dt^2} + R\frac{di}{dt} + \frac{1}{C}i = 0. \tag{b}$$

In equations (a) and (b) the voltage E is constant. In many cases, however, E is a variable function of the time, most often represented by a sine, cosine, or combinations of these functions. In general we may represent it as $E = f(t)$. In this case the right member of (a) is $f(t)$ and the derivative with respect to t is not zero but $f'(t)$. This gives, instead of (b),

$$L\frac{d^2i}{dt^2} + R\frac{di}{dt} + \frac{1}{C}i = f'(t). \tag{c}$$

Our previous differential equations contain only first derivatives and are said to be equations of the *first order;* the last two equations above contain a second derivative and are said to be differential equations of the *second order*. In general, the order of a differential equation is that of the highest derivative which appears in it. Differential equations of the first and second order are of the greatest importance in physics and engineering and the subject of differential equations forms a large branch of mathematics.   The methods of integration which we have studied in this book are not sufficient to solve the second order equation above and the reader who is interested in the subject must refer to books on differential equations.

82. Effective Value of Alternating Current.   *An alternating current of electricity is not steady but continually rises, falls and reverses, twice becoming zero and twice rising to a maximum (in opposite directions) in a complete cycle of changes.   Find the equivalent steady current or "effective" alternating current when the maximum value for the cycle is known.*

Solution.   If I is the maximum value for the cycle, i is the instantaneous value at any time t, and f is the frequency in cycles per second, then i is given by the formula

$$i = I \sin (2\pi ft).$$

The quantity $(2\pi ft)$ is called the *phase angle* of the cycle and designated by θ.   Hence

$$i = I \sin \theta. \qquad (a)$$

In a complete cycle, $\theta = 360$ degrees, and the graph of this equation is given in Fig. 41.   The curve OABCD is repeated for every cycle and I

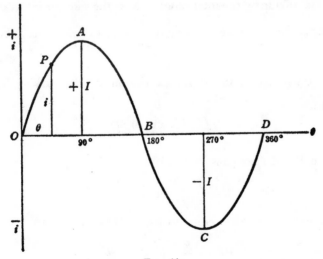

FIG. 41.

is the maximum height.   The ordinate of any point as P on the curve is the instantaneous value of the current i corresponding to the phase θ.

The loop OAB of the curve is exactly equal and opposite to the loop BCD, and the current generates the same amount of heat in the negative half cycle as in the positive half cycle.   The effective current is, therefore, the same for each half cycle, and for the whole cycle, and as long as the current continues to flow.   We shall, therefore, calculate the effective value for the half cycle OAB.

At any instant when the phase is θ the current is i, as shown at P. The heating effect is not proportional to the current itself but to the square of the current, and when the current i flows in a resistance R the energy converted into heat per second is Ri^2.   But by (a) $i = I \sin \theta$. Hence the heat generated per second is

$$Ri^2 = R(I \sin \theta)^2 = RI^2 \sin^2 \theta.$$

The average value of this heat energy for all values of θ over the entire cycle is the same as would be generated by the equivalent steady current.   If we let I_e represent this equivalent or *effective* current, then the heat generated by it per second is RI_e^2 and this must be equal to the *average* of $RI^2 \sin^2 \theta$ over the cycle, or half cycle.

In order to find the average value of any quantity over a certain range, we sum up all its values in that range (or *integrate* it over that range) and divide by the total range.   Now the sum (or integral) over a half cycle, 0 to 180°, or 0 to π, is $\int_0^\pi$ and therefore the average is $\dfrac{\int_0^\pi}{\pi}$.   The average of the above expression for the varying heat is, therefore,

$$\int_0^\pi \frac{(RI^2 \sin^2 \theta)\, d\theta}{\pi}$$

and since R and I are constants, this is,

$$\frac{RI^2}{\pi} \int_0^\pi \sin^2 \theta \, d\theta.$$

As explained above this is equal to RI_e^2.   Hence,

$$RI_e^2 = \frac{RI^2}{\pi} \int_0^\pi \sin^2 \theta \, d\theta.$$

Dividing this by R we have

$$I_e^2 = \frac{I^2}{\pi} \int_0^\pi \sin^2 \theta \, d\theta, \tag{b}$$

and in order to determine the value of the effective current I_e we must integrate $\int \sin^2 \theta \, d\theta$.

In order to transform this for integration, we use the relation from trigonometry which states that $\cos 2\theta = 1 - 2 \sin^2 \theta$, and hence,

$$\sin^2 \theta = \tfrac{1}{2}(1 - \cos 2\theta).$$

$$\therefore \int \sin^2 \theta \, d\theta = \tfrac{1}{2} \int (1 - \cos 2\theta) \, d\theta = \tfrac{1}{2} \int d\theta - \tfrac{1}{2} \int \cos 2\theta \, d\theta$$

$$= \tfrac{1}{2}\theta - \tfrac{1}{2} \int \cos 2\theta \, d\theta.$$

Now, $\tfrac{1}{2} \int \cos 2\theta \, d\theta = \tfrac{1}{4} \int 2 \cos 2\theta \, d\theta = \tfrac{1}{4} \int \cos 2\theta \cdot 2 \, d\theta$

$$= \tfrac{1}{4} \int \cos 2\theta \cdot d(2\theta) = \tfrac{1}{4} \sin 2\theta.$$

Hence, $\int \sin^2 \theta \, d\theta = \tfrac{1}{2}\theta - \tfrac{1}{4} \sin 2\theta = \tfrac{1}{2}(\theta - \tfrac{1}{2} \sin 2\theta)$, and hence by (b),

$$I_e^2 = \frac{I^2}{2\pi} \left[\theta - \frac{1}{2} \sin 2\theta \right]_0^\pi = \frac{I^2}{2\pi} \left[\left(\pi - \frac{1}{2} \sin 2\pi \right) - \left(0 - \frac{1}{2} \sin 0 \right) \right]$$

$$= \frac{I^2}{2\pi} \left(\pi - \frac{1}{2} \sin 360° \right) = \frac{I^2}{2\pi} (\pi).$$

$$\therefore \quad I_e^2 = \frac{I^2}{2}$$

and,

$$I_e = \frac{I}{\sqrt{2}} = .7071.$$

Thus the effective value of an alternating current is about .71 or a little less than three fourths of the maximum current.   Similarly for voltage in an A.C. circuit

$$E_e = .707E, \tag{d}$$

where E is the maximum value of the voltage in the cycle.

Alternating-current meters indicate effective values of current and voltage. When reading the effective value on a meter, it is well to remember that twice in each cycle the current (or voltage) rises to a value which according to equations (c), (d) is $1/.707 = 1.41$ or nearly one and one-half times as great as the meter reading. Thus, if an A.C. meter reads 100 volts on a 60-cycle circuit the shock received by placing the hands across the circuit would be 141 volts 120 times a second.

83. Acceleration Against Air Resistance. *The air resistance to a moving object is, within certain limits, proportional to its speed, increasing as the speed increases. If the weight of a car and the tractive force exerted by its engine are known, find the speed at a specified time after starting from rest on a level road.*

Solution. At any specified moment let t be the time since starting and and let the velocity at that moment be v. Then, since the force of the wind resistance f_w is proportional to the speed, we have,

$$f_w = kv, \tag{a}$$

where k is the proportionality constant. If F represents the pulling force of the engine, then when this is opposed to the wind resistance the net force effective in accelerating the car is $f_a = F - f_w$, and hence, according to equation (a),

$$f_a = F - kv. \tag{b}$$

One of the fundamental principles or laws of mechanics is that the force used to accelerate an object equals the product of the mass by the acceleration, and mass is weight divided by g (acceleration of gravity), W/g, while the acceleration is as we have seen, dv/dt. The accelerating force is therefore

$$f_a = \frac{W}{g} \cdot \frac{dv}{dt}. \tag{c}$$

Comparing (b) and (c) we have at once

$$\frac{W}{g} \cdot \frac{dv}{dt} = F - kv,$$

or,

$$\frac{W}{g} \cdot \frac{dv}{dt} + kv = F. \tag{d}$$

If this differential equation is compared with equation (b) of article 79 or (a) of article 80 it is seen to be of the same form. Integration therefore, give as the solution

$$v = \frac{F}{k}\left(1 - e^{-\frac{kg}{W}t}\right). \tag{e}$$

Since k, g, W, F are known, this formula will give the speed v at any time t after starting, or give the time t required to bring the car up to any speed v.

Formula (e) will also apply to the case of a large object falling through the air from a height, as, for example, a person jumping from an aeroplane with an opened parachute which is held back from falling freely by the resistance of the air.   In this case the downward force F exerted to accelerate the person and parachute is their combined weight and k as in (e) expresses the air resistance, which is found by experiment with exposed surfaces in different wind velocities.

As an illustration of this use of the formula (e), suppose a 135-pound person jumps with a 25-pound parachute which opens at once and has a surface area of 200 square feet exposed to the air.   If the air pressure on the parachute is two pounds per square foot when falling at 30 feet per second, and the person reaches the ground in 30 seconds, with what velocity does he strike the ground?

In this case the pulling force is the weight and $F = W = 135 + 25 = 160$ pounds, and $g = 32$ (ft./sec.) per second.   The force of air resistance when $v = 30$ ft./sec. is $f_w = 2 \times 200 = 400$ pounds and hence by equation (a), $400 = k(30)$ so that $k = 400/30 = 40/3$.   The time is $t = 30$ seconds, and we have still to find v.

Summarizing the data, we have $F = 160$, $W = 160$, $k = 40/3$, $g = 32$, $t = 30$.   Using these values in formula (e), we get

$$\frac{F}{k} = \frac{160}{(40/3)} = \frac{3 \times 160}{40} = 12,$$

$$\frac{kg}{W}t = \frac{(40/3) \times 32 \times 30}{160} = 80.$$

$$v = 12(1 - e^{-80}).$$

From a table of exponentials or by logarithms we find that e^{-80} is an extremely small fraction so that $1 - e^{-80} = 1$ (approximately).   Therefore, the person lands with a velocity of

$$v = 12 \text{ feet per second.}$$

In order to find the velocity with which he would strike if the parachute should fail to open we must know the height from which he jumps and this can be found by combining some of the equations of this article with some of those of article 76 dealing with freely falling objects. Equation (c) of article 76 states that when the acceleration is g and the time of fall t the velocity is gt. This is true for *any* acceleration a; if an object is steadily accelerated with the acceleration a the final velocity is $v = at$. Hence, if the time and velocity are known, we can find the acceleration

$$a = v/t.$$

In this case the person falls for 30 seconds and attains a velocity of 12 feet per second. The acceleration is therefore,

$$a = 12/30 = 2/5 \text{ (ft./sec.) per sec.}$$

Also by equation (e), article 76, for any acceleration a the distance fallen through is $s = \frac{1}{2}at^2$. Hence, in this case, the height is

$$s = \frac{1}{2}(2/5)(30)^2 = 180 \text{ feet.}$$

If he had fallen *freely* (without the open parachute) for 180 feet the full force of gravity unchecked by the air resistance would have been effective and the acceleration 32 instead of 2/5. Then, by the formula $s = \frac{1}{2}gt^2$ he would have fallen for a time

$$t = \sqrt{\frac{2s}{g}} = \frac{1}{4}\sqrt{s} = \frac{1}{4}\sqrt{180} = 3.36 \text{ seconds}$$

instead of 30 seconds, and his velocity would have been

$$v = gt = 32 \times 3.36 = 107.5 \text{ ft./sec.}$$

or,

$$v = 107.5/88 = 1.22 \text{ mi./min.}$$

84. Time of Swing of a Pendulum.   Suppose it is required *to find the time required by a pendulum or any suspended body to swing from its extreme position in one direction to that in the other* when the total distance of swing is not very great.

Solution.   In Fig. 42 let O be the point of suspension, OA the length L of the pendulum or suspension and A, B its extreme positions. The pendulum bob, or other swinging body, will then swing back and forth in the circular arc ACB, C being the lowest point and vertically under O. Draw the lines OB and OC and let α represent the angle AOC, which is equal to the angle BOC, and let m be the mass of the pendulum bob, the suspending cord or rod being very light as compared with the bob.

As the bob swings toward the left and reaches A it stops, reverses and returns toward C with its velocity accelerated from A to C. When it reaches some point P it makes an angle POC $= \theta$ with the

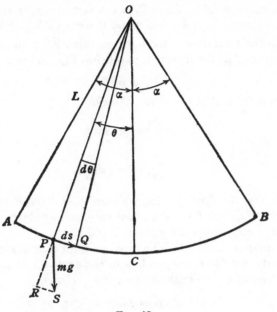

FIG. 42.

center line OC and in the next instant of time dt it will swing through the small angle POQ $= d\theta$ and reach Q, passing over the very small distance PQ $= ds$. Its velocity during this interval is ds/dt and its acceleration is $\dfrac{d^2s}{dt^2}$ and as the line OC and the point C are the reference line and point, the distance s from C is *decreasing* so that ds/dt and $\dfrac{d^2s}{dt^2}$ are negative.

The force required to accelerate an object is the product of its mass by the acceleration.   Therefore, the force in this case is

$$f = -m\frac{d^2s}{dt^2}.\qquad\text{(a)}$$

This accelerating force is due to the weight of the body which is its mass multiplied by the acceleration of gravity, mg. We can, therefore, represent the weight by the vertical downward pointing arrow marked mg in Fig. 42. That part of the weight effective in the direction of PQ is found by extending OP to R and drawing RS parallel to PQ, which is perpendicular to OP. Then PRS is a right triangle and the angle RPS $= \theta$. Hence $\overline{RS}/\overline{PS} = \sin \theta$ and therefore $\overline{RS} = \overline{PS} \sin \theta$. But when $\overline{PS}$ is the total downward force mg, then $\overline{RS}$ is the effective accelerating force f in the direction parallel to PQ. Hence

$$f = mg \sin \theta. \tag{b}$$

By equations (a) and (b) we have, therefore,

$$-m \frac{d^2s}{dt^2} = mg \sin \theta.$$

$$\therefore \quad \frac{d^2s}{dt^2} = -g \sin \theta. \tag{c}$$

In order to find the time of swing we must solve this differential equation in terms of s, t and θ and then solve the resulting algebraic equation for t when proper values are given to s and θ.

Before equation (c) can be integrated we must express s in terms of θ. When s is the arc CP corresponding to the angle θ and L, the length of OP, is the radius, we know that $s = L\theta$. Hence

$$ds/dt = L(d\theta/dt)$$

and

$$\frac{d^2s}{dt^2} = L \frac{d^2\theta}{dt^2}. \tag{d}$$

Substituting (d) in equation (c) and dividing by L we get finally as the equation to be integrated,

$$\frac{d^2\theta}{dt^2} = -\frac{g}{L} \sin \theta. \tag{e}$$

We have here a second derivative to be integrated, or a second order differential equation to solve, and must use a new device. In order to transform the equation for integration, let us multiply both sides by $2\left(\dfrac{d\theta}{dt}\right) dt$. This gives

$$2\left(\frac{d\theta}{dt}\right) \cdot \left(\frac{d^2\theta}{dt^2}\right) dt = -\frac{2g}{L} \sin \theta \cdot \left(\frac{d\theta}{dt}\right) dt. \tag{f}$$

But, $\dfrac{d^2\theta}{dt^2} = \dfrac{d}{dt}\left(\dfrac{d\theta}{dt}\right)$, therefore

$$2\left(\frac{d\theta}{dt}\right)\left(\frac{d^2\theta}{dt^2}\right)dt = 2\left(\frac{d\theta}{dt}\right)\cdot\frac{d}{dt}\left(\frac{d\theta}{dt}\right)dt = 2\left(\frac{d\theta}{dt}\right)\cdot d\left(\frac{d\theta}{dt}\right) = d\left[\left(\frac{d\theta}{dt}\right)^2\right].$$

Also, on the right side of (f), the dt cancels in the numerator and denominator. Equation (f), therefore, becomes

$$d\left[\left(\frac{d\theta}{dt}\right)^2\right] = -\frac{2g}{L}\sin\theta\,d\theta,$$

which can now be integrated. Integrating,

$$\int d\left[\left(\frac{d\theta}{dt}\right)^2\right] = -\frac{2g}{L}\int\sin\theta\,d\theta.$$

$$\therefore \quad \left(\frac{d\theta}{dt}\right)^2 = \frac{2g}{L}\cos\theta + C. \tag{g}$$

In order to determine the integration constant C, note that at the point A just as the pendulum is stopped to reverse, the angle is α and $d\theta$ is zero. That is $d\theta/dt = 0$ and $\theta = \alpha$. These values in (g) give

$$0 = (2g/L)\cos\alpha + C, \quad \therefore \quad C = -(2g/L)\cos\alpha,$$

and equation (g) becomes

$$\left(\frac{d\theta}{dt}\right)^2 = \frac{2g}{L}(\cos\theta - \cos\alpha).$$

Taking the square root of both sides of this equation,

$$\frac{d\theta}{dt} = \sqrt{\frac{2g}{L}}\cdot\sqrt{\cos\theta - \cos\alpha}.$$

$$\therefore \quad \frac{d\theta}{\sqrt{\cos\theta - \cos\alpha}} = \sqrt{\frac{2g}{L}}\,dt.$$

It is the *time* we wish to find, so let us transpose this equation, getting

$$dt = \sqrt{\frac{L}{2g}}\frac{d\theta}{\sqrt{\cos\theta - \cos\alpha}}.$$

Integrating,

$$\int dt = \sqrt{\frac{L}{2g}}\int\frac{d\theta}{\sqrt{\cos\theta - \cos\alpha}}. \tag{h}$$

The expression $\displaystyle\int\frac{d\theta}{\sqrt{\cos\theta - \cos\alpha}}$ cannot be integrated, but by remembering that the swing is to be small, that is, α and θ are small, we can transform $\cos \theta - \cos \alpha$ so that it is integrable. From trigonometry we have the relation

$$\cos \theta = 1 - 2 \sin^2 \left(\tfrac{1}{2}\theta\right) \tag{i}$$

and when an angle is very small (5° or less), if it is expressed in radian measure it is very nearly equal to its sine. Therefore, in the present case $\sin \left(\tfrac{1}{2}\theta\right) = \tfrac{1}{2}\theta$ and $\sin^2\left(\tfrac{1}{2}\theta\right) \left(\tfrac{1}{2}\theta\right)^2 = \tfrac{1}{4}\theta^2$. Hence,

$$\cos \theta = 1 - \tfrac{1}{2}\theta^2.$$

Similarly,

$$\cos \alpha = 1 - \tfrac{1}{2}\alpha^2.$$

$$\therefore \quad \cos \theta - \cos \alpha = \left(1 - \tfrac{1}{2}\theta^2\right) - \left(1 - \tfrac{1}{2}\alpha^2\right) = -\frac{\theta^2}{2} + \frac{\alpha^2}{2} = \frac{\alpha^2 - \theta^2}{2}$$

and

$$\sqrt{\cos \theta - \cos \alpha} = \frac{\sqrt{\alpha^2 - \theta^2}}{\sqrt{2}}.$$

This value of the radical when used in equation (h) gives

$$\int dt = \sqrt{\frac{L}{g}} \int \frac{d\theta}{\sqrt{\alpha^2 - \theta^2}}.$$

Now t and θ are corresponding values of the time and angle when the pendulum is in any position P, Fig. 42. At the extreme position A, therefore, $\theta = \alpha$ and t is half the time of a complete swing, $\tfrac{1}{2}T$. We, therefore, integrate from $\theta = 0$ to $\theta = \alpha$ and from $t = 0$ to $t = \tfrac{1}{2}T$. Our integrals then become

$$\int^{\frac{T}{2}} dt = \sqrt{\frac{L}{g}} \int_0^a \frac{d\theta}{\sqrt{\alpha^2 - \theta^2}}.$$

Integrating by formula (XXIII),

$$\int \frac{d\theta}{\sqrt{\alpha^2 - \theta^2}} = \sin^{-1}\left(\frac{\theta}{\alpha}\right).$$

$$\therefore \quad [t]_0^{\frac{T}{2}} = \sqrt{\frac{L}{g}}\left[\sin^{-1}\left(\frac{\theta}{\alpha}\right)\right]_0^a.$$

$$\therefore \quad \frac{T}{2} = \sqrt{\frac{L}{g}}\left[\sin^{-1}\left(\frac{\alpha}{\alpha}\right) - \sin^{-1}\left(\frac{0}{\alpha}\right)\right] = \sqrt{\frac{L}{g}}\left(\sin^{-1}1\right) = \sqrt{\frac{L}{g}}\left(\frac{\pi}{2}\right).$$

$$\therefore \quad T = \pi \sqrt{\frac{L}{g}} \tag{j}$$

is the time required for one swing of a pendulum of length L.   Since $g = 32.2$ (ft./sec.) per sec., T is in seconds and L in feet.

As an illustration let us find the length of the pendulum of a jeweler's clock which beats seconds.   In this case $T = 1$ and formula (j) gives

$$1 = \pi \sqrt{\frac{L}{32.2}} \quad \text{or} \quad L = \frac{32.2}{\pi^2}.$$

$$\therefore \quad L = 3.26 \text{ ft.} = 1 \text{ meter.}$$

Thus the meter pendulum beats seconds.

In order to beat any desired time, as in clocks or musicians' time-beaters, the pendulum, according to (j), must be of length

$$L = \left(\frac{g}{\pi^2}\right) T^2.$$

85. Surface of a Liquid in a Rotating Vessel.   *When a vessel containing a liquid is rotated rapidly the surface of the liquid rises at the edges and sinks in the center.   Find the shape of the surface when the speed of rotation is known.*

Solution.   The shape of the surface is due to the centrifugal force resulting from the rotation, which drives the liquid toward the edge of the vessel, and when the vessel and the liquid in it are rotating steadily and the curved surface of the liquid is at rest every particle of liquid in the surface is in equilibrium with the force of gravity and the centrifugal force; and the resultant of these two is perpendicular to the surface, for if it were inclined the particle would move downward, but it is not moving downward.

In Fig. 43 let ABCD represent the body of liquid and the curve KOL the curved surface of the liquid.   In order to determine the shape of the curve we must find its equation.   Draw the vertical OY and the horizontal OX through the lowest point O of the surface curve.   Then, if P represents any point on the curve the horizontal MP = x is its abscissa and PQ = y is its ordinate.   We are to find the equation giving the relation between x and y.

If the vessel rotates n times per second and the mass of a particle at P is m the centrifugal force acting on it is $(mv^2) \div \overline{MP}$ or mv^2/x where v is its velocity.   But when x is the radius of the circle in which it rotates n times per second the velocity is $2\pi x \cdot n$, and hence the centrifugal force is $m(2\pi xn)^2/x = m(2\pi n)^2 x$.   This horizontal force may be represented by the arrow PF and we can write

$$\overline{PF} = m(2\pi n)^2 x. \tag{a}$$

The force of gravity acting on the particle of mass m is mg, where $g = 32.2$ is the acceleration of gravity, and is vertically downward. Representing this force by the arrow PG, we have

$$\overline{PG} = mg. \tag{b}$$

The resultant of these two forces is PR and this is perpendicular to the surface at P, as seen above. If we draw the tangent PT then PT is perpendicular to PR.

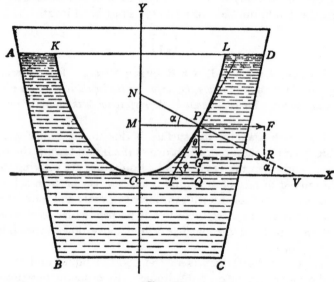

FIG. 43.

Now produce PR in both directions to meet OX at V and OY at N. Then the triangles GPR and MNP are similar right triangles and therefore we have the proportion between the sides:

$$\overline{NM} : \overline{MP} = \overline{PG} : \overline{GR}. \tag{c}$$

But $\overline{GR} = \overline{PF}$ and $\overline{MP} = x$. Using these values and substituting (a) and (b) for $\overline{PF}$ and $\overline{PG}$ in (c), the proportion becomes

$$\overline{NM} : x = mg : m(2\pi n)^2 x.$$

Solving this proportion for $\overline{NM}$ we get

$$\overline{NM} = \frac{g}{(2\pi n)^2}. \tag{d}$$

Since MP is parallel to OV, the angle MPN equals the angle OVP. If we call this angle α, then in the right triangle MPN we have $\tan \alpha = \overline{NM}/\overline{MP}$. Using in this the value of $\overline{NM}$ from (d) and remembering that $\overline{MP} = x$ it becomes

$$\tan \alpha = \frac{\left[\dfrac{g}{(2\pi n)^2}\right]}{x}. \tag{e}$$

We have also in the right triangle TVP, $\tan \alpha = \cot \phi = 1/\tan \phi$. But since PT is tangent to the curve at P we have $\tan \phi = dy/dx$. Therefore

$$\tan \alpha = \frac{1}{\left(\dfrac{dy}{dx}\right)}.$$

Comparing this with equation (e), we see that

$$\frac{1}{\left(\dfrac{dy}{dx}\right)} = \frac{\left[\dfrac{g}{(2\pi n)^2}\right]}{x} \quad \text{or} \quad \frac{dy}{dx} = \frac{(2\pi n)^2}{g}x.$$

$$\therefore \quad dy = \frac{(2\pi n)^2}{g} x \, dx.$$

Integrating this differential relation to get the relation between x and y,

$$\int dy = \frac{(2\pi n)^2}{g} \int x \, dx.$$

$$\therefore \quad y = \frac{(2\pi n)^2}{2g} x^2 + C. \tag{f}$$

Since P is *any* point on the curve, this equation holds good for any point on the curve, and hence at O.   But at O, x is zero and y is zero. Putting these two values for x and y in (f), we find that C is zero and (f) becomes simply

$$y = \frac{(2\pi n)^2}{2g} x^2 = \left(\frac{2\pi^2 n^2}{g}\right) x^2.$$

Now π, n, g are all constants and hence $\left(\dfrac{2\pi^2 n^2}{g}\right)$ is a constant.   If we let this constant be represented by the single letter a, then our equation becomes

$$y = ax^2 \tag{g}$$

and this is the equation of the curve KOPL in Fig. 43, since it is the relation between x and y for any point on the curve.

Since the constant $a = 2\pi^2 n^2/g$ is known when the revolutions per second are known, the equation (g) gives the distance y above the lowest point O for any point on the surface whose distance from the axis of rotation OY is x.   Since it applies to any point on the surface, it therefore applies to all and gives the shape of the entire surface when the liquid is rotated about OY at the specified speed.

86. The Parabola. If we compare equation (g) above with equation (43), article 33, and the curve KOPL of Fig. 43 with the curve KOP of Fig. 15, article 33, we see at once that they are the same curve and equation with $a = 1$ in (43).   The curve KOP or KOPL is called a *parabola*.

The constant a may have different values and the corresponding curves will then be either steeper and narrower or broader and flatter, but the general shape remains the same.   Several different forms of parabola are shown in Fig. 44 with the equation for each.

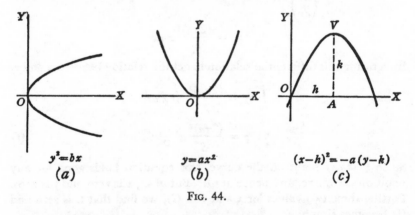

$$y^2 = bx$$
(a)

$$y = ax^2$$
(b)

$$(x-h)^2 = -a(y-k)$$
(c)

FIG. 44.

The point O or V in either case is called the *vertex* of the parabola and may be at O or any other point, depending on the choice of lines for the coordinate axes.   The curve may open upward or downward, to the right or left.   This is indicated by the sign of the multiplying constant and by the appearance of the exponent 2 with either the x or the y. Thus in Fig. 44 (b) the constant a is preceded by a plus sign and the curve opens upward while it is reversed in Fig. 44 (c). The symmetrical center line through the vertex is called the *axis* of the parabola. In Fig. 44 (a) the y is squared and the OX axis is the axis of the parabola; in (b) the x is squared and OY is the axis.

The curve Fig. 44 (c) is seen to be the same as the path of the projectile shown in Fig. 38. In Fig. 38 P is the vertex and PF is the axis of the parabola.

The equation of the parabola of Fig. 38 is equation (g), article 77,

$$y = (\tan \alpha)x - \left(\frac{g}{2V^2 \cos^2 \alpha}\right)x^2,$$

and by the algebraic process of completing the square on the right side of this equation (as in solving quadratic equations), and transposing, it takes the form

$$\left[x - \left(\frac{V^2 \sin 2\alpha}{2g}\right)\right]^2 = -\left(\frac{2V^2 \cos^2 \alpha}{g}\right)\left[y - \left(\frac{V^2 \sin^2 \alpha}{2g}\right)\right]$$

and this is seen to be the same as

$$[x - h]^2 = -a[y - k],$$

which is the same as the equation of Fig. 44 (c), the constants a, h, k replacing the expressions in parentheses.

The parabola is an extremely important curve in science and engineering. Thus the cross section of the reflector of an automobile or locomotive headlight is a parabola, so also is the suspended loaded cable of a suspension bridge. We have already seen that the path of a projectile is a parabola; this can easily be made visible by turning a water hose upward at an angle and adjusting the nozzle so that the water issues in a small, clear jet and does not spray. The particles or drops of water are projectiles and the steady stream of them makes a continuous parabola. Other examples of the parabola are, the path of a non-returning comet in the heavens, certain architectural arches, vertical curves on railway and highway bridges, the water surface investigated in the preceding article, etc.

Although the subject of this article belongs to analytical geometry rather than to calculus, it is of such interest and importance that we give here the general definition of a parabola and the derivation of its equation. The definition is, "A parabola is the curve traced out by a point which moves so that its distances from a fixed point and a fixed line are always equal." To derive the equation, let F be the fixed point in Fig. 45, DD' the fixed line and P the moving point. Then according to the definition of the curve the distance $\overline{PF} = \overline{PM}$.

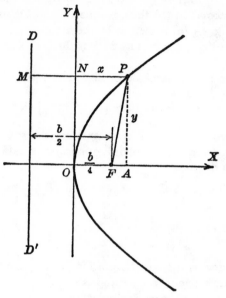

Fig. 45.

Draw FX through F perpendicular to DD' and OY parallel to DD' midway between F and DD', and draw PA perpendicular to OX. Then the coordinates of P are $NP = OA = x$ and $AP = y$. In order to find the equation we must find the relation between x and y.

Let the distance of the fixed point F from the fixed line DD' be designated by $\frac{1}{2}b$. Then

$$\overline{OF} = \tfrac{1}{2}(\tfrac{1}{2}b) = \tfrac{1}{4}b \quad \text{and} \quad \overline{AF} = \overline{OA} - \overline{OF} = x - \tfrac{1}{4}b.$$

In the right triangle PAF the distance $\overline{PF} = \sqrt{\overline{AF^2} + \overline{AP^2}}$.

$$\therefore \quad \overline{PF} = \sqrt{(x - \tfrac{1}{4}b)^2 + y^2}. \tag{a}$$

Also the distance $\overline{PM} = \overline{PN} + \overline{NM}$, or

$$\overline{PM} = x + \tfrac{1}{4}b. \tag{b}$$

Now to trace out the parabola the point P moves so that always

$$\overline{PF} = \overline{PM}.$$

According to equations (a) and (b) therefore

$$\sqrt{(x - \tfrac{1}{4}b)^2 + y^2} = (x + \tfrac{1}{4}b).$$

Squaring,

$$(x - \tfrac{1}{4}b)^2 + y^2 = (x + \tfrac{1}{4}b)^2.$$

Carrying out the indicated squaring and simplifying, this becomes

$$y^2 = bx, \tag{c}$$

which is the same as in Fig. 44 (a).   Since this is the relation between the abscissa and ordinate of a point moving according to the definition of the parabola, it is the equation of the parabola.

The fixed point F is called the *focus* of the parabola and the fixed line DD' is the *directrix*.   As already stated, O is the *vertex* and OX is the *axis*.

In the case of a non-returning comet moving in a parabolic path the sun is at the focus and in the parabolic reflector the light is placed at the focus.

The parabola possesses the remarkable property, as can be shown from its equation, that if a line from the focus to a point P on the curve, as FP in Fig. 46, meets at P a line parallel to the axis, as PL, then the line PK which is perpendicular to the curve at P makes *angle KPF equal angle KPL*.   This is the law of reflection of light or sound, hence, if a source of light or sound is placed at F a sound wave or a ray of

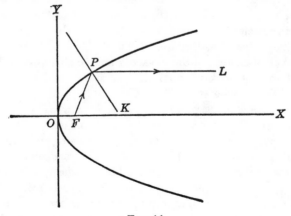

FIG. 46.

light from F is reflected at P along the line PL *parallel to the axis.*
Since P is any point, this is true for all points and so all rays from F are
reflected parallel to the axis.   It is for this reason that parabolic re-
flectors are suitable for searchlights and the light is placed at the focus:
the rays form a direct parallel beam and are not scattered.

The parabolic reflector is a bowl-shaped vessel of which the cross-
section through its center-line or axis is a parabola.   One method of
forming the parabolic inner surface is to fill a vessel such as that of
Fig. 43 with a fine cement or plaster and mount it on a vertical shaft
with the line OY as axis; the vessel is then rotated at the proper speed
so that the parabolic surface takes the desired shape.   The rotation
continues at the same speed until the plaster hardens, and the surface
then holds the parabolic shape permanently.   It is then coated with
glass or metal or other reflecting material and mounted in its holder
with a supporting backing.

It is to be noted that the equation of the parabola contains only the
first power of one variable and the second power (or second and first)
of the other variable.   This is characteristic of many formulas and
equations in science, but we shall give only one example here in addition to those already cited above.   If an electric current of strength I
amperes flows through a resistance R ohms the power lost in heat for
any value of I is

$$P = RI^2$$

watts, and this is seen to be of the same form as the equation of Fig.
44 (b).   The curve of Fig. 44 (b) and also that of Fig. 15 which was
discussed in articles 33, 64 and 67 are therefore the graph of the electrical resistance heat loss formula.

For further discussion of the parabola and related curves the reader
is referred to any book on analytical geometry and to Wentworth's
"Plane and Solid Geometry," Book IX.

87. Work Done by Expanding Gas or Steam.   The characteristic
property of gases is that when a definite mass of gas occupies any
volume at any pressure the product of the pressure and volume is the
same for all pressures and volumes for the same mass of gas so long as
the temperature remains constant.   Thus if the gas is compressed
from a larger to a smaller volume the pressure is increased as the
volume is decreased, but the product of the new pressure and volume
is the same as that of the old.   In this case work is done *on* the gas, as

in an air compressor.   If the gas expands from a smaller to a larger volume the pressure decreases and work is done *by* the gas (in pushing back the restraining wall or piston), as in a steam engine.   Let us *find the work done in either case*.   It will be simpler to handle the case of the expanding gas.

Solution.   The law that the product of pressure and volume remains constant as pressure p and volume v vary is expressed by writing

$$pv = k \qquad \text{(a)}$$

where k is the characteristic constant, which has the same value for all pure gases and can be found by experiment.

The graph of equation (a) is shown in Fig. 47 (a) and is called a

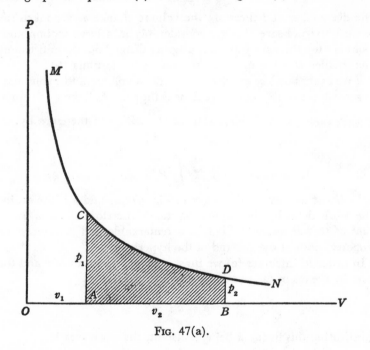

FIG. 47(a).

hyperbola.   It extends indefinitely in both directions.   We now proceed to determine the formula for the work done by the expanding gas and in so doing we discover a remarkable property of the curve of Fig. 47 (a).

Suppose the gas is contained in a cylinder and held at a volume v

under a pressure p by a very light frictionless piston of area A. When the gas expands a small amount it pushes the piston back a small distance ds against the pressure and does an amount of work dW. The work done is equal to the force multiplied by the distance through which it acts. The force exerted by the gas on the piston is the pressure p times the area A of the piston, that is, pA, and the distance through which it acts is ds. The work done is, therefore,

$$dW = pA \cdot ds.$$

But $A \cdot ds$ is the change in volume which occurs when the piston moves through the distance ds, and this is dv. Hence,

$$dW = p \, dv. \tag{b}$$

The distance ds and therefore the volume change dv are taken small, because the pressure changes considerably in a large motion and we could not use the same p as the multiplier throughout the entire motion. The smaller ds is taken, the more exact is the formula (b).

If we next allow the gas to expand from a volume v_1 to a considerably larger volume v_2, the total work done is the sum of all the small amounts of work such as dW. Hence $W = \int_{v_1}^{v_2} dW$, and therefore by equation (b)

$$W = \int_{v_1}^{v_2} p \, dv. \tag{c}$$

But v is the abscissa of the curve in Fig. 47 (a) and p is the ordinate. The work done by the expanding gas is therefore the *area under the graph of the gas equation!* This is a remarkable and extremely useful property, both of the gas and of the hyperbola.

In order to integrate (c) we must express p in terms of v and this is done by means of the relation (a) which gives

$$p = k \cdot \frac{1}{v}.$$

Substituting this in (c), k being constant, the **work-area** is

$$W = k \int_{v_1}^{v_2} \frac{dv}{v}.$$

This integral is the familiar and fundamental form $\int \frac{dv}{v} = \log_e v$. Therefore,

$$W = k[\log_e v]_{v_1}^{v_2} = k(\log_e v_2 - \log_e v_1).$$

$$\therefore \quad W = k \log_e \left(\frac{v_2}{v_1}\right) \tag{d}$$

is the work done by the gas or steam in expanding from volume v_1 to volume v_2.

The area given by formula (c) or (d) is shown shaded in Fig. 47 (a), where OA $= v_1$, OB $= v_2$ and AC $= p_1$, BD $= p_2$ are the corresponding pressures. In a steam sengine a definite amount of steam is let into the cylinder and expands against the piston, doing work. After expansion, most of the steam is exhausted and on the back stroke new steam is let in and partially compressed, so that some of the work done by the steam is spent again and the curve takes the form of Fig. 47 (b). This is called the engine *indicator card* and the shaded area LMNQRL is measured to obtain the work done per stroke. The strokes per minute being known for the speed of the engine, the total power developed by the engine can then be calculated.

In Fig. 47 (b) the volume v_2 is the total volume of the cylinder and v_1 is the clearance volume into which steam is let when the piston is all the way in. The steam is said to be expanded along the hyperbola MN and compressed along QRL. In an actual indicator card the

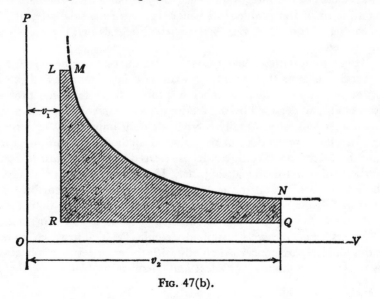

FIG. 47(b).

corners at L, M, N, Q, R are never actually square, on account of the impossibility of opening and closing the cylinder valves with absolute suddenness.

If the gas in the cylinder is *compressed* from the larger volume v_2 to the smaller volume v_1 the work given by formula (c) or (d) is done *on* the gas, as is the case in an air compressor. In either case when v_1 and v_2 or the *expansion ratio* or *compression ratio* v_2/v_1 are given, the work can be calculated for any gas or vapor for which the characteristic constant k is known.

As stated above, the corners of the work area are not square in an actual indicator card of an engine or compressor and because of unavoidable influences in the operation of the machine the calculation of the work or work area by the theoretical formula may be difficult. In this case it is measured by an integrating instrument called the *planimeter*, which performs the integration mechanically and automatically.

88. Remarks on the Application of Integration. The problems discussed in this chapter are not selected for either their practical value or theoretical interest alone, but for both, and in the solution and discussions the aim has been to show not only the simple applications and details of the integration itself, but, in particular, the nature and conditions of the problem in which the integrals originate, and the meaning of the connection between the integrals and the phenomena described.

It has been thought better to solve and discuss at some length a few typical problems than simply to write out the statement and integral in each case for a large number of problems. A few such problems thoroughly understood by the reader are more instructive than a large number which are formally solved but only imperfectly understood.

For other applications of the calculus methods, both in integration and in differentiation, the reader is referred to the standard textbooks on scientific and technical subjects. If further discussion is desired of both the calculus and the technical and scientific problems in which it is useful, the present author knows of no better books than Perry's "Calculus for Engineers," Graham's "Calculus for Engineering Students," and Bisacre's "Applied Calculus." The first two, as their titles indicate, are concerned chiefly with the engineering applications, while the third discusses physical and chemical problems as well. An elementary book of particular interest to chemists is Daniels' "Mathematical Preparation for Physical Chemistry."

It has no doubt been noticed that in the technical applications and natural phenomena discussed in the preceding articles, the logarithmic and exponential functions occur very frequently. This is true in general in natural phenomena and is in itself a natural phenomenon, which is discussed in the next chapter.

89. Problems for Solution.

The following problems require only the slightest technical knowledge, and most of what is required is explained in the solutions in this chapter. The symbols should be selected and the equations or formulas developed and set up before any numerical values are used. The methods and procedures explained and illustrated in this chapter will suffice for the solution of all of the problems.

1. With what velocity will a stone strike the ground if (a) dropped from the roof of a building 100 ft. high?   (b) thrown downward from the top of the same building with a speed of 60 ft. per second?

2. A stone dropped from a balloon which was rising at the rate of 20 ft. per second, reached the ground in 6 seconds.   (a) How high was the balloon when the stone was dropped?   (b) With what velocity did the stone strike the ground?

3. A projectile thrown from the top of a tower at an angle of 45° above the horizontal strikes the ground in 5 sec. at a horizontal distance from the foot of the tower equal to its height. Find the height of the tower.

4. The temperature of a liquid in a room of temperature 20° is observed to be 70°, and five minutes later to be 60°.   Assume Newton's Law of cooling: The rate of cooling (decrease of temperature) is proportional to the difference of the temperatures of the liquid and the room; and find the temperature of the liquid 30 minutes after the first observation.

5. A boat moving in still water is subject to a retardation (resistance) proportional to its velocity at any instant.   Let V be the velocity at the instant the power is shut off, and find the velocity v at the end of t seconds after the power is shut off.

6. A small mass sliding on a certain inclined plane is subject to an acceleration downward along the plane of 4 ft. per sec. per sec.   If it is started upward from the bottom of the incline with a velocity of 6 ft. per sec. find the distance it moves in t sec.   How far will it go before stopping and starting to slide back?

7. If the inclined plane in Prob. 6 is 20 ft. long, find the necessary initial speed in order that the mass may just reach the top.

8. Atmospheric pressure p at points above the earth's surface as a function of the altitude h above sea-level changes according to the "population" or

"compound-interest" law of article 75. If $p = 15$ lb./sq. in. at sea-level ($h = 0$), and 10 lb. when $h = 10,000$ ft., find p when (a) $h = 5000$, and (b) $h = 15,000$ ft.

9. The velocity of a chemical reaction in which the amount of the reacting substance changed or transformed in time t is x, is the time rate of change of x, dx/dt. In a so-called "first order reaction," when a is the initial amount of the substance, the amount remaining unchanged at the end of time t is $a - x$. The law of the first order reaction is: The rate of transformation is proportional to the amount present unchanged. The constant of proportionality is called the "velocity constant." Find the formula for the velocity constant (k).

10. The chemical reaction known as the "inversion of raw sugar" is a first order reaction. If 1000 lb. of raw sugar is reduced to 800 lb. in 10 hours, how much will remain unchanged at the end of 24 hours?

11. In the process known as "washing down a solution" water is run into a tank containing the solution (salt or acid) for the purpose of reducing its strength. If s is the quantity of salt (or) acid in the tank at any time, and x is the amount of water which has then run through, find the rate of decrease of s with respect to $x(ds/dx)$, and from this result find s.

12. In Prob. 11, if the total volume (salt and water) is $v = 10,000$ gal., how much water must be run through to wash down 50% of the salt?

13. In an electrical condenser which is discharging electricity, the time rate of change of voltage (E) is proportional to E, and E decreases with the time (t). If the proportionality constant is $k = 40$, find the time required for E to fall to 10% of its original value.

14. "Building up" a salt (or acid) solution by adding salt (or acid) while the volume remains constant (v) leads to the equation $dy/dx = (v - y)/v$, where y is the amount of salt (or acid) in the tank at any moment and x is the amount added from the beginning. Solve this equation for y in terms of v and x.

15. The acceleration of an object moving through the air at a high speed v near the earth's surface is in a particular case $32 - (v^2/8)$ ft. per sec. per sec. If such an object falls to the ground in one minute from rest, find the height from which it falls.

16. Radium decomposes at a rate proportional to the amount present at any instant. If half the original amount disappears in 1600 years, find the amount remaining at the end of 100 years.

17. A drop of liquid in air evaporates at a rate proportional to the area of its surface. Find the radius of the drop as a function of the time.

18. In a certain first order chemical reaction, the amount of reagent remaining unchanged at the end of one hour after the reaction begins is 48 lb., and at the end of 3 hours, 27 lb. What was the amount at the start?

19. A vertical tank has a slight leak at the bottom. Assuming that the water escapes at a rate proportional to the pressure (depth multiplied by density) and that 5% escapes the first day, how long will it be before the tank is half empty?

20. If the population of a certain region changes according to the natural law from 40,000 to 90,000 in 40 years, what will it be at the end of 20 years more?

21. The amount of light absorbed in passing through a thin sheet of water is proportional to the amount (A) falling on the surface and also to the thickness (t) of the sheet.   If one third the light is absorbed in penetrating 10 ft., how much of the original amount will reach a depth of 60 ft.?

22. The rate of change of vapor pressure of a liquid with respect to temperature is proportional to the vapor pressure and inversely proportional to the square of the absolute temperature.   If the vapor pressure of water is 4.58 mm. of mercury (barometer) at zero Centigrade (273° absolute) and 31.8 mm. at 30°C., determine the equation expressing the vapor pressure (P) as a function of the absolute temperature (T).

23. Disregarding air resistance and the earth's curvature and rotation, find the range, greatest height reached, and time of flight of a rifle bullet which is fired with a muzzle velocity of 1500 ft. per sec. at an elevation of 30 degrees. Also write and simplify the equation of its trajectory.

24. In the parabolic trajectory of article 77, which has the form of Fig. 44(c), the height is $k = 18$ mi. and the range is $2h = 72$ mi.   Also, at 0, $x = 0$ and $y = 0$.   Find the constant a, and the area under the curve above the horizontal.

25. In order to form a parabolic reflector (article 86) by the method of article 85, to have a depth of 6 in. and a diameter of 6 in. across the open end of the bowl-shaped figure, what speed of rotation is required?

Chapter 18

THE NATURAL LAW
OF GROWTH AND THE NUMBER e

90. Introduction. In Chapter 10 we have seen that a certain number designated by the symbol e is the number which when used as the base for logarithms gives the simplest formula for the differential or derivative of a logarithm and in Chapter 11 we have seen that the integral of a reciprocal, which is the one exception to the general rule for integration of powers, furnishes a new definition of a logarithm, provided that e is used as the base.

We have also seen in Chapters 10 and 13 that the xth power of e is its own derivative as well as its own integral and in Chapter 17 that logarithmic and exponential functions based on e occur frequently in applied mathematics. Truly this number e is a remarkable number and ranks with the geometrical number π in importance.

In the present chapter we investigate the origin and significance of the number e, calculate its value, and determine some of the properties of both e and e^x.

91. The Natural Law of Growth. We have all heard such sayings as "riches beget riches" and "more calls for more" and the like, but we probably have not stopped to think of the meaning back of such statements. If we do stop to think, however, the statements are seen to be reasonable and their meaning is clear.

As an illustration, consider the increase of money at compound interest ("riches beget riches"). Taking any specific amount of money as the principal or starting amount and allowing it to increase at a definite rate (of interest) without withdrawing any of the accumulation, it is seen at once that in each interest period there is added a definite fraction of what was present at the beginning of the period, and so at the beginning of the next period there is a larger principal sum than at the beginning of the last. In the new period the same fraction but a larger amount is added, and so on, and the increase is increasingly rapid.   It is to be noted, however, that the increase is always the *same fraction* of what is already present, it is the amount present which increases.

Again, consider the normal rate of increase of population (article 75).   If there is no national calamity and no immigration and all may expect to attain to the allotted "three-score-and-ten" then each individual will ordinarily see more births than deaths so that each generation will have more parents and there will be more children; even though the average size of families may not change, *there will be more families*.   As in the case of compound interest, it is not the rate that changes but the *amount present* which is productive.

The gist of both examples just given may be expressed in the statement that in natural processes the increase in any period is proportional to the quantity present.   Thus the mass of a growing tree or plant is increased by the formation of new cells by all those already present, the consumption of fuel by fire increases as more heat is generated and more heat is generated as the consumption of fuel increases; the process depends on the amount of heat present (assuming unlimited fuel).

On the other hand, if we think of the cooling of a hot object we see at once that at first when it is hot it will throw off more heat in a certain time because it is much hotter than its surroundings, which can thus absorb rapidly, but later it will have less heat and so be cooler, and will not throw off so much in the same length of time, the amount radiated again being proportional to the amount present.   Similarly in a chemical reaction or the discharge of an electrified object (such as an electrical condenser, article 79) as the reaction or discharge continues (without being replenished) there is less active substance present at each successive instant and therefore less activity and the decrease of active substance is less, again being *proportional to the amount present*.

The citation of further examples would only be repetition.   It is seen that the process of change in nature, whether increase or decrease, the natural law of growth (or decay) when not influenced by external or artificial means, can be stated as follows: *The rate of change of any active quantity at any time is proportional to the amount present at that time.*

92. Mathematical Expression of the Law of Growth.   The idea developed in the preceding article is easily expressed in mathematical form.   At the beginning of any process such as those given as examples there will be a definite amount of the thing under consideration and in every succeeding unit of time, that is, each second, minute, day, year, etc., a definite portion or fraction of the amount present at that time is added or removed.

If A is the amount present at any particular time t and a is the *fraction* undergoing change in one unit of time, then the rate of change is the total amount undergoing change in the unit of time and is the fraction multiplied by the amount present, that is, it is $a \times A$. But the rate of change is dA/dt. Therefore

$$\frac{dA}{dt} = aA. \tag{89}$$

This equation states that the rate of change dA/dt is *proportional to the amount present*, A, and that a is the constant of proportionality. In order to find the amount present at any specified time t the differential equation (89) is to be solved for A in terms of t.

Transforming (89), there results

$$\frac{dA}{A} = a \, dt.$$

Integrating,

$$\log_e A = at + C.$$

In order to determine the integration constant C we make use of the fact that at the beginning of the process, when $t = 0$, there is a definite known amount A_0 present. Using the values $t = 0$ and $A = A_0$ in the last equation and solving for C we get $C = \log_e A_0$, and the equation becomes

$$\log_e A = at + \log_e A_0.$$

Transposing, $\log_e A - \log_e A_0 = at$ or $\log_e \left(\dfrac{A}{A_0} \right) = at$. Hence, $\dfrac{A}{A_0} = e^{at}$ and finally,

$$A = A_0 e^{at}.$$

In the case where the change is a decrease dA/dt is negative and hence $dA/dt = -aA$ which gives $A = A_0 e^{-at}$.

The form of the equation which applies to either growth or decay can therefore be written

$$A = A_0 e^{\pm at}, \tag{90}$$

and this states that in the natural law of growth (or decay) the amount A of the substance or quantity undergoing change which is present at any time t after the process starts is an exponential function of the time with the original amount as the coefficient of the exponential, the fraction added (or removed) per unit of time as the coefficient of the variable exponent t, and *the number e as the base!*

93. The Compound Interest Formula. In order to express in another form the idea expressed in the preceding articles we take the case of money at compound interest as a typical example of the natural law of growth and find a formula which shall express it in a form suitable for calculation. Since what is true for one rate will be true for any rate and principal, we take one dollar as principal and the convenient decimal fraction one tenth (10%) as rate and set ourselves the problem of finding a formula for the amount at the end of ten years when the interest is compounded at different intervals.

If the dollar is invested at simple interest one tenth of the original amount will be added each year and at the end of the ten years one dollar interest is added and the amount is two dollars.

At compound interest, however, if it is compounded annually the amount added the first year is one tenth as in simple interest and at the end of the year the amount is 1.1 dollars or $1 + \frac{1}{10}$. At the beginning of the second year this amount is taken as the new principal, so that there is 1.1 dollars to increase and the amount added in the second year is one tenth of this or 11 cents so that the total is 1.21 dollars. That is, there is at the end of the second year $1 + \frac{1}{10}$ times as much present as at the beginning, and since at the beginning of the second year there was already $1 + \frac{1}{10}$ times as much as at the beginning of the first year, the total at the end of the second year is the 1.1 times 1.1 or

$$(1 + \tfrac{1}{10})(1 + \tfrac{1}{10}) \text{ dollars.}$$

At the beginning of the third year the amount present is therefore $[(1 + \frac{1}{10})(1 + \frac{1}{10})]$ and one tenth of this is added during the year so that at the end of the third year there is $(1 + \frac{1}{10})$ times that amount, that is,

$$(1 + \tfrac{1}{10})[(1 + \tfrac{1}{10})(1 + \tfrac{1}{10})].$$

Similarly, the amount at the beginning of the fourth year is that present at the end of the third year and at the end of the fourth year $(1 + \frac{1}{10})$ that amount, or $(1 + \frac{1}{10})(1 + \frac{1}{10})(1 + \frac{1}{10})(1 + \frac{1}{10})$.

It is now easy to see what the amount will be at the end of the fifth or any other year, and to save further repetition we will simply write out the amounts for several years. Thus there will be at the end of the

1st year, $(1 + \tfrac{1}{10})$ $= (1 + \tfrac{1}{10})^1$

2nd year, $(1 + \tfrac{1}{10})(1 + \tfrac{1}{10})$ $= (1 + \tfrac{1}{10})^2$

3rd year, $(1 + \tfrac{1}{10})(1 + \tfrac{1}{10})(1 + \tfrac{1}{10})$ $= (1 + \tfrac{1}{10})^3$

4th year, $(1 + \tfrac{1}{10})(1 + \tfrac{1}{10})(1 + \tfrac{1}{10})(1 + \tfrac{1}{10})$ $= (1 + \tfrac{1}{10})^4$

5th year, $(1 + \tfrac{1}{10})(1 + \tfrac{1}{10})(1 + \tfrac{1}{10})(1 + \tfrac{1}{10})(1 + \tfrac{1}{10}) = (1 + \tfrac{1}{10})^5$

Without writing any other amounts we now see at once that at the end of the tenth year the amount in dollars will be

$$A = (1 + \tfrac{1}{10})^{10} \tag{91}$$

when the principal is 1 dollar and the compounding at 10% is annual, that is, when there are ten interest periods. If this is calculated by means of logarithms, expressed as 1.1^{10}, it will be found to be 2.594 dollars.

Suppose now that the rate remains the same and the total time to run is the same, but that the compounding is semi-annual. Then in each half year only 5% or $\tfrac{1}{20}$ is added, but in the 10 years there will be 20 periods. Now formula (91) gives the amount when there are ten periods and $\tfrac{1}{10}$ the amount present at the beginning of each period is added during the period. By the same process that was used to obtain formula (91) it is easily seen that at the end of the 20 periods when $\tfrac{1}{20}$ is the fraction added in each period the amount will be

$$A = (1 + \tfrac{1}{20})^{20}. \tag{92}$$

Similarly, if the rate is 10% and the time is 10 years, but the compounding is every four months, the amount added in each period will be $\tfrac{1}{30}$, there will be 30 periods and the amount will be

$$A = (1 + \tfrac{1}{30})^{30}. \tag{93}$$

Similarly, if the interest were compounded 10 times a year and 1% or $\tfrac{1}{100}$ added each time, there would be 100 interest periods in the 10 years and the amount would be

$$A = (1 + \tfrac{1}{100})^{100}. \tag{94}$$

Let us look back now to the case of simple interest and see how it is to be formulated. Since the interest is allowed to accumulate steadily throughout the entire period on the same principal without starting anew at stated periods with a new principal composed of the amount present at the end of the preceding period, there is only one interest period ten years long and in that time not simply a fraction of a dollar but one whole dollar is added, the amount is evidently

$$A = (1 + \tfrac{1}{1})^1 = 2 \tag{95}$$

dollars, as we saw before.

Suppose now the money runs for five years at simple interest and is then re-invested with accumulated interest for another five years at simple interest at the same rate. At 10% the interest in the first five years is .5 dollar, and in the next five years the interest will be half the new principal of 1.5 dollars, or .75 dollar, and the total at the end of the five years is therefore 2.25 dollars. Thus there have been two interest periods and $\frac{1}{2}$ was added each time; the amount is therefore

$$A = (1 + \tfrac{1}{2})^2 = 2.25. \tag{96}$$

By comparison of the formulas (91) to (96) we can now write out the formula for the amount in the ten years at 10% with any number of interest periods and by the use of logarithms can calculate the amount.

The question now arises as to whether the amount will increase indefinitely if the interest periods are taken shorter and shorter. In other words, if the compounding takes place continuously, every instant, so that the fraction added each instant is indefinitely small but the number of periods is indefinitely large, what is the amount? The answer to this question is given in the table below.

In this table the number of interest periods is represented by n, the fraction of the amount present at the beginning of any period which is added during that period is $1/n$ and the amount is

$$A = \left(1 + \frac{1}{n}\right)^n. \tag{97}$$

n	$\left(1 + \dfrac{1}{n}\right)^n$	A
1	$(1 + \frac{1}{1})^1$	2.000
2	$(1 + \frac{1}{2})^2$	2.250
3	$(1 + \frac{1}{3})^3$	2.369
5	$(1 + \frac{1}{5})^5$	2.489
10	$(1 + \frac{1}{10})^{10}$	2.594
20	$(1 + \frac{1}{20})^{20}$	2.653
30	$(1 + \frac{1}{30})^{30}$	2.672
50	$(1 + \frac{1}{50})^{50}$	2.691
100	$(1 + \frac{1}{100})^{100}$	2.7048
1000	$(1 + .001)^{1000}$	2.7170
10000	$(1 + .0001)^{10000}$	2.7182
100000	$(1 + .00001)^{100000}$	2.71828
......	....	....
∞	....	$2.7182818284+ \ldots$

Since n may be any number, the formula (97) will represent the amount for any rate or time when the rate for the period is related to the number of interest periods in the same way as in our present problem. For any other rate, the formula would be correspondingly modified but it would still be exponential with the number of interest periods as the exponent. For any other principal than one dollar the amount as given for one dollar is simply multiplied by the principal.

Our problem of the continuous compounding resolves itself, therefore, into the calculation of A in formula (97) when n is very large. In the table this is indicated by taking n larger and larger and letting ∞ (infinity) represent the largest possible value of n. Then it is seen that our formulas (95), (96), (91), (92), (93), (94), respectively, are the formulas for $n = 1, 2, 10, 20, 30, 100$, (97) is the general formula and for $n = \infty$, $A = 2.7182818284 + \cdots$.

In the table it is easily seen how the value of A is computed for any specified finite value of n, but the method of computing A when $n = \infty$ is not immediately obvious, for if we simply put $n = \infty$ in the formula $\left(1 + \dfrac{1}{n}\right)^n$, since $1/\infty = 0$, we would have

$$A_\infty = (1 + 0)^\infty = 1^\infty = 1,$$

instead of $2.7182818284 + \cdots$.

We cannot therefore simply substitute $n = \infty$ in the formula but must look upon A_∞ as a sort of *limiting value* to which A becomes more and more nearly equal as n *approaches* infinity, that is, the value which A has when n differs from infinity by an amount which may be as small as we please, or when n becomes indefinitely large. This value is called the "limit of $\left(1 + \dfrac{1}{n}\right)^n$ as n approaches infinity" and is written

$$A_\infty = \operatorname*{Lim}_{n \to \infty} \left(1 + \frac{1}{n}\right)^n. \tag{98}$$

The method of calculating this value is given in a later article. We shall find that this is the *formula for calculating the number e;* it is no doubt already observed that the last value in the table above is the *same as that given for e* in article 46.

94. A New Definition of the Derivative. We have just seen that the amount of one dollar at compound interest when the number of interest periods is very large, or as we said *approaches infinity*, is found by calculating the *limiting value* of the quantity A as given by formula

(97). The calculation of ultimate or limiting values of a dependent variable as the independent variable approaches some specific value, infinity, zero, or any other, is a process which is frequently useful and it can be made the basis of the definition of the derivative. This method of defining and calculating derivatives can be made very short and concise in many cases if one is thoroughly familiar with the notion of limits, but for ordinary purposes the best definition is the one which is based on rates, because the idea of a rate is a natural one and the calculation of differentials and derivatives on that basis requires only a knowledge of simple algebra and trigonometry.

In order to get a different view of the significance of the number e, however, and to find a method of calculating its value, we shall give here the limit definition of the derivative and show how it is used, and then apply it to find the derivative of $\log_b x$.

If we have any function of x denoted by $f(x)$ and wish to find the derived function, or derivative, $f'(x)$ by the limit method we proceed as follows: First suppose x to be increased by an amount h; then instead of x we will have $x + h$ and instead of the original function $f(x)$ we will have a new function $f(x + h)$, the difference between the new and old function being $f(x + h) - f(x)$. This difference divided by the quantity h is called the *difference quotient;* it is, therefore,

$$\frac{f(x + h) - f(x)}{h}. \tag{99}$$

The added quantity h is next supposed to become smaller and smaller indefinitely so that it *approaches zero as a limit;* then $x + h$ becomes more and more nearly equal to x, $f(x + h)$ approaches the original function $f(x)$, and the difference $f(x + h) - f(x)$ also becomes smaller and smaller indefinitely. The difference *quotient* (99) will then become some new function of x, and it turns out that this new function is the *derived function* $f'(x)$, which is the derivative dy/dx if $y = f(x)$, and with which we have already become acquainted in connection with the differentials and rates of functions.

We can therefore say that the derived function or derivative is the *limiting value which the difference quotient approaches as the difference h in the value of x approaches toward zero* indefinitely. Using the same form of notation that we used to indicate the final limiting value of the compound interest formula, we can write for the derivative of any function of x with respect to x,

$$f'(x) = \underset{h \to 0}{\text{Lim}} \left[\frac{f(x + h) - f(x)}{h} \right]. \tag{100}$$

As an illustration of the working of this method let us apply it to the simple function x^2 whose derivative we already know to be $2x$. We write

$$f(x) = x^2;$$

then

$$f(x + h) = (x + h)^2$$

$$= x^2 + 2xh + h^2$$

$$f(x + h) - f(x) = (x^2 + 2xh + h^2) - x^2 = 2xh + h^2 = (2x + h)h.$$

$$\therefore \quad \frac{f(x + h) - f(x)}{h} = 2x + h.$$

$$\therefore \quad f'(x) = \underset{h \to 0}{\text{Lim}} \, (2x + h).$$

Now as h becomes indefinitely nearly equal to zero, $2x + h$ becomes simply $2x$. Therefore the limiting value is

$$f'(x) = 2x,$$

which is the value of the derivative.

This method can be used to find the derivative of any expression and all our derivative or differential formulas could have been obtained in this way, but it is a formal and somewhat artificial method and does not seem to connect so directly and naturally with problems involving varying quantities as does the ordinary notion of rates.

For the purpose already stated, however, we shall apply this method to find the derivative of log x to any base and thereby obtain a formula for the calculation of e.

95. Derivative of Log$_b$ x. *Formula for e.* To find the derivative of $\log_b x$ by the limit method when b is any base we have

$$f(x) = \log_b x;$$

then

$$f(x + h) = \log_b (x + h).$$

$$f(x + h) - f(x) = \log_b (x + h) - \log_b x$$

$$= \log_b \left(\frac{x + h}{x} \right) = \log_b \left(1 + \frac{h}{x} \right).$$

$$\therefore \quad f(x + h) - f(x) = \log_b \left[1 + \frac{1}{\left(\dfrac{x}{h}\right)} \right].$$

Now, according to formula (51), article 44,

$$m \log_b M = \log_b (M^m)$$

$$\therefore \quad \frac{1}{m} (m \log_b M) = \frac{1}{m} \cdot \log_b (M^m).$$

Therefore, we can write our last formula above as

$$f(x + h) - f(x) = \frac{1}{\left(\dfrac{x}{h}\right)} \left\{ \left(\dfrac{x}{h}\right) \log_b \left[1 + \frac{1}{\left(\dfrac{x}{h}\right)} \right] \right\}$$

$$= \left(\frac{h}{x}\right) \cdot \log_b \left[1 + \frac{1}{\left(\dfrac{x}{h}\right)} \right]^{\left(\frac{x}{h}\right)},$$

and by dividing both sides (first and last member) of this equation by h we get for the difference quotient

$$\frac{f(x + h) - f(x)}{h} = \frac{1}{x} \cdot \log_b \left[1 + \frac{1}{\left(\dfrac{x}{h}\right)} \right]^{\left(\frac{x}{h}\right)}.$$

Hence the derivative, by formula (100) above, is

$$f'(x) = \operatorname*{Lim}_{h \to 0} \left\{ \frac{1}{x} \cdot \log_b \left[1 + \frac{1}{\left(\dfrac{x}{h}\right)} \right]^{\left(\frac{x}{h}\right)} \right\}.$$

For simplicity in writing, let us put $\dfrac{x}{h} = n$. Then as h becomes extremely small, n, which is (x/h), becomes extremely large and $n \to \infty$ means the same as $h \to 0$. Thus we can write our formula for the derivative

$$f'(x) = \operatorname*{Lim}_{n \to \infty} \left\{ \frac{1}{x} \cdot \log_b \left[1 + \frac{1}{n} \right]^n \right\}.$$

The only part of this expression affected by n becoming very large is that part which contains n, that is, the expression in square brackets. Therefore, instead of indicating the limit of the entire expression we can write the formula so as to indicate the limit of that part only. The formula then becomes finally

$$f'(x) = \frac{1}{x} \cdot \log_b \left[\operatorname*{Lim}_{n \to \infty} \left(1 + \frac{1}{n} \right)^n \right].$$

But

$$f'(x) = \frac{1}{x} \cdot \log_b e, \tag{101}$$

according to our derivative formula (T), article 50. Comparing the two formulas, we see at once that

$$e = \operatorname*{Lim}_{n \to \infty} \left(1 + \frac{1}{n} \right)^n. \tag{102}$$

In deriving formula (101) in article 45 the number e appeared simply and naturally as a particular constant involved in the usual algebraic formulas connected with logarithms, but we there had no way of calculating its value. We now have a formula, (102), which will give us the value of e when the indicated computation is carried out. This expression for e appeared in the derivation of the derivative of $\log_b x$ by the limit method in the same place in the formula and in the same general connection, but in an entirely different manner, but since we have seen that the limit method gives the same result as the differential method in the case of x^2 and can also readily obtain the same result by the two methods for any other function, the two formulas for the derivative of $\log_b x$ must be the same and, therefore, formula (102) is the true expression for the value of the number e. We shall carry out the computation of the numerical value of e by formula (102) in the next article.

By comparing (102) with formula (98) it is seen at once that the two are the same. The amount of one dollar at compound interest when the compounding is continuous, is therefore *equal to e!* When we calculate e in the next article, therefore, we also compute the last value in the table of article 92.

96. Calculation of e. In algebra we learn that any binomial $a + b$ can be raised to any power n by the following formula:

$$(a + b)^n = a^n + na^{n-1}b + \frac{n(n-1)}{1 \cdot 2} a^{n-2}b^2$$
$$+ \frac{n(n-1)(n-2)}{1 \cdot 2 \cdot 3} a^{n-3}b^3 + \cdots,$$

and for a specified integer value of the exponent n the number of terms in the expression is equal to $n + 1$, but when n is not specified the series runs on indefinitely.

If $a = 1$, the formula becomes

$$(1 + b)^n = 1 + nb + \frac{n(n - 1)}{1 \cdot 2} b^2 + \frac{n(n - 1)(n - 2)}{1 \cdot 2 \cdot 3} b^3 + \cdots, \quad (103)$$

and in this expansion of $(1 + b)^n$, the term b may be any number whatever. If we use $b = \frac{1}{n}$ we have $\left(1 + \frac{1}{n}\right)^n$, which is the expression whose limiting value is e when n becomes indefinitely large.

When $b = \frac{1}{n}$, we only have to replace b by $\frac{1}{n}$ in the expansion (103); this gives us, when the expression is simplified,

$$\left(1 + \frac{1}{n}\right)^n = 1 + n\left(\frac{1}{n}\right) + \frac{n(n - 1)}{1 \cdot 2}\left(\frac{1}{n}\right)^2$$
$$+ \frac{n(n - 1)(n - 2)}{1 \cdot 2 \cdot 3}\left(\frac{1}{n}\right)^3 + \cdots$$

$$= 1 + 1 + \frac{n(n - 1)}{1 \cdot 2}\frac{1}{n^2} + \frac{n(n - 1)(n - 2)}{1 \cdot 2 \cdot 3}\frac{1}{n^3} + \cdots$$

$$= 1 + 1 + \frac{n^2\left(1 - \frac{1}{n}\right)}{1 \cdot 2n^2} + \frac{n^3\left(1 - \frac{1}{n}\right)\left(1 - \frac{2}{n}\right)}{1 \cdot 2 \cdot 3n^3} + \cdots,$$

$$\therefore \quad \left(1 + \frac{1}{n}\right)^n = 2 + \frac{1 - \frac{1}{n}}{1 \cdot 2} + \frac{\left(1 - \frac{1}{n}\right)\left(1 - \frac{2}{n}\right)}{1 \cdot 2 \cdot 3}$$
$$+ \frac{\left(1 - \frac{1}{n}\right)\left(1 - \frac{2}{n}\right)\left(1 - \frac{3}{n}\right)}{1 \cdot 2 \cdot 3 \cdot 4} + \cdots.$$

In order to calculate e, according to formula (102), we must find the value of each of the terms in this series when n is infinitely large. If we make n infinitely large, each of the fractions with n in the denominator, $1/n$, $2/n$, $3/n$, etc., becomes infinitely small, or zero, and each of the expressions in parentheses, $1 - 1/n$, $1 - 2/n$, $1 - 3/n$, etc., is simply 1. The series then gives

$$e = \operatorname*{Lim}_{n \to \infty} \left(1 + \frac{1}{n}\right)^n = 2 + \frac{1}{1 \cdot 2} + \frac{1}{1 \cdot 2 \cdot 3} + \frac{1}{1 \cdot 2 \cdot 3 \cdot 4}$$
$$+ \frac{1}{1 \cdot 2 \cdot 3 \cdot 4 \cdot 5} + \cdots$$

and, therefore, we have finally,

$$e = 2 + \tfrac{1}{2} + \tfrac{1}{6} + \tfrac{1}{24} + \tfrac{1}{120} + \tfrac{1}{720} + \tfrac{1}{5040} + \cdots.$$

In order to carry out the remainder of the computation, it is only necessary to express the fractions as decimals to as many decimal places as desired and add the results. This work can be greatly simplified by noting a peculiarity of the series of fractions. The term $\frac{1}{6}$ is the term $\frac{1}{2}$ divided by 3, $\frac{1}{24}$ is $\frac{1}{6}$ divided by 4, $\frac{1}{120}$ is $\frac{1}{24}$ divided by 5, $\frac{1}{720}$ is $\frac{1}{120}$ divided by 6, and so on. If, therefore, we write $\frac{1}{2} = .5$, divide this by 3 to as many decimal places as desired, divide the resulting quotient by 4, divide this quotient by 5, the next by 6, the next by 7, etc., and add the results, the value of e is obtained immediately. The division can be carried out by short division and the quotients written one under the other successively as far as it is desired to carry the computation. In this way the result is found to be

$$e = 2.7182818284$$

to ten decimal places. Like π this number is an unending decimal and its computation has been carried to a great number of places. Since the computation is, however, so simple, it has not excited the interest which the computation of π has aroused. We shall find in a later article that π can also be computed by means of a series of fractions, and that this series is based on e.

97. Calculation of e^x for Any Value of x. Having examined briefly the relation of the number e to natural processes and found its numerical value, we next develop a formula for calculating the value of the exponential e^x when the value of x is given. This formula will be used later in the calculation of logarithms and the trigonometric functions as well as π.

We have seen that when n is very large $e = \left(1 + \frac{1}{n}\right)^n$, therefore, when n is still very large, $e^x = \left[\left(1 + \frac{1}{n}\right)^n\right]^x = \left(1 + \frac{1}{n}\right)^{nx}$. Expressed in symbols the full statement is

$$e^x = \operatorname*{Lim}_{n \to \infty} \left(1 + \frac{1}{n}\right)^{nx}, \tag{104}$$

and we can expand $\left(1 + \dfrac{1}{n}\right)^{nx}$ in a series formula, each term of which can be computed when n is infinite.

Using the binomial expansion, as it was used in the preceding article, we get,

$$\left(1 + \frac{1}{n}\right)^{nx} = 1 + nx\left(\frac{1}{n}\right) + \frac{nx(nx-1)}{1\cdot 2}\left(\frac{1}{n}\right)^2$$
$$+ \frac{nx(nx-1)(nx-2)}{1\cdot 2\cdot 3}\left(\frac{1}{n}\right)^3 + \cdots$$

and again taking out n as a factor in each of the parentheses,

$$\left(1 + \frac{1}{n}\right)^{nx} = 1 + nx\left(\frac{1}{n}\right) + \frac{n^2 x\left(x - \dfrac{1}{n}\right)}{1\cdot 2n^2}$$
$$+ \frac{n^3 x\left(x - \dfrac{1}{n}\right)\left(x - \dfrac{2}{n}\right)}{1\cdot 2\cdot 3n^3} + \cdots$$
$$= 1 + x + \frac{x\left(x - \dfrac{1}{n}\right)}{1\cdot 2} + \frac{x\left(x - \dfrac{1}{n}\right)\left(x - \dfrac{2}{n}\right)}{1\cdot 2\cdot 3} + \cdots.$$

To find the limit which this expression approaches when n is infinite, we again suppose that in each of the fractions $1/n$, $2/n$, etc., n is infinite; each fraction then becomes zero and each term in parentheses is simply x. The limit of the entire series is then,

$$\operatorname*{Lim}_{n\to\infty}\left(1 + \frac{1}{n}\right)^{nx} = 1 + x + \frac{x\cdot x}{1\cdot 2} + \frac{x\cdot x\cdot x}{1\cdot 2\cdot 3} + \frac{x\cdot x\cdot x\cdot x}{1\cdot 2\cdot 3\cdot 4} + \cdots,$$

and hence, according to (104),

$$e^x = 1 + x + \frac{x^2}{1\cdot 2} + \frac{x^3}{1\cdot 2\cdot 3} + \frac{x^4}{1\cdot 2\cdot 3\cdot 4} + \frac{x^5}{1\cdot 2\cdot 3\cdot 4\cdot 5} + \cdots. \quad (105)$$

The form of the terms is very simple and it is an easy matter to write out and compute as many as may be desired. The calculation can be carried out in a manner similar to that used in the computation of e.

98. Derivative of e^x. By formula (105),

$$e^x = 1 + x + \frac{x^2}{1\cdot 2} + \frac{x^3}{1\cdot 2\cdot 3} + \frac{x^4}{1\cdot 2\cdot 3\cdot 4} + \frac{x^5}{1\cdot 2\cdot 3\cdot 4\cdot 5} + \cdots. \quad (105)$$

Differentiating this expression as a sum of separate terms, we get

$$\frac{d(e^x)}{dx} = 0 + 1 + \frac{2x}{1\cdot2} + \frac{3x^2}{1\cdot2\cdot3} + \frac{4x^3}{1\cdot2\cdot3\cdot4} + \frac{5x^4}{1\cdot2\cdot3\cdot4\cdot5} + \cdots$$

$$\therefore \quad \frac{d(e^x)}{dx} = 1 + x + \frac{x^2}{1\cdot2} + \frac{x^3}{1\cdot2\cdot3} + \frac{x^4}{1\cdot2\cdot3\cdot4} + \cdots.$$

Comparing this series with the original series (105), we see that they are precisely the same term by term. Therefore,

$$\frac{d(e^x)}{dx} = e^x$$

which is the already familiar expression for the derivative of e^x.

Chapter 19

SOLUTION OF
DIFFERENTIAL EQUATIONS

Differential equations were introduced in Chapter 17, and examples were given of a few simple types. Many other types are known; in fact, they are of great importance in science, engineering, and other fields of applied mathematics. The first step, therefore, in a broader discussion of them is to define briefly the different kinds that may exist.

99. Classification of Differential Equations. An *ordinary differential equation* is a relation involving at most two variables, that is, an independent variable and a dependent variable, together with their derivatives (or differential coefficients), which may be of any order. Thus, if x is the independent variable and y the dependent one, the most general ordinary differential equation is of the form

$$f\left(x, y, \frac{dy}{dx}, \frac{d^2y}{dx^2}, \ldots, \frac{d^ny}{dx^n}\right) \tag{1}$$

where the use of the symbol f denotes that the terms in x, y, $\frac{dy}{dx}$, etc. may be present, in any of their functions, as $3x$, x^2, y^3, $\cos x$, $\log y$, $4\left(\frac{dy}{dx}\right)^2$, $x\left(\frac{d^3y}{dx^3}\right)^4$, etc. It also implies that the equation may contain constant terms, such as 6, $4a$, $9b^2$, etc.

Note that the definition of an ordinary differential equation restricts it to a single independent variable and a single dependent variable. A differential equation involving two or more independent variables, one dependent variable, and partial derivatives (or differential coefficients) of the dependent variable with respect to one or more of the independent variables is a *partial differential equation*.

A differential equation involving two or more dependent variables, one independent variable (which may not appear explicitly) and derivatives (or differential coefficients) with respect to the independent variable is a *total differential equation*.

243

The *order* of a differential equation is the order of the highest derivative present.   Thus, $\frac{d^2y}{dx^2}$ is a second order derivative, $\frac{d^3z}{dx^3}$ is a third order derivative, etc.

The *degree* of a differential equation is the exponent of the derivative of highest order after the equation is completely rationalized and cleared of fractions.   To illustrate the determination of degree, let us consider the following equations

$$\frac{d^2y}{dx^2} + \left(\frac{dy}{dx}\right)^3 + 4y - 6 = 0 \tag{2}$$

$$\frac{d^2y}{dx^2} = \sqrt{1 + y^2 + \frac{dy}{dx}} \tag{3}$$

Equation (2) is of the first degree, because the exponent of its highest derivative, $\frac{d^2y}{dx^2}$ is 1.   On the other hand, equation (3) is of the second degree because in order to rationalize it (remove the radical sign) we must square both sides of the equation, obtaining the derivative of highest order, $\frac{d^2y}{dx^2}$ as its square, $\left(\frac{d^2y}{dx^2}\right)^2$.

An ordinary differential equation is *linear* if it is of the first degree in the dependent variable and all of its derivatives, and if they do not occur as products of each other.

The *solution of a differential equation* is any functional relation between the variables, free from derivatives, which causes the differential equation to become an identity.   The process of finding a solution usually involves algebraic manipulation, such as the rationalization operation cited above, followed by integration.   For the latter operation, tables of integrals are useful.   Moreover, there are also tables of solutions of differential equations.*   In order to use any of these aids effectively, however, some knowledge of the procedure in solving the simpler equations is necessary.

100. First Order Equations with One Variable Missing.   In other words, the equation is a function of only the derivative and the one variable.   The method of solution is first to solve for the derivative and then to integrate.

---

* G. M. Murphy, "Ordinary Differential Equations and Their Solutions," Van Nostrand, 1960.

Example.   Solve the equation

$$\left(\frac{dy}{dx}\right)^2 + 2\frac{dy}{dx} + x = 0. \tag{4}$$

Transposing
$$\left(\frac{dy}{dx}\right)^2 + 2\frac{dy}{dx} = -x.$$

Adding 1,
$$\left(\frac{dy}{dx}\right)^2 + 2\frac{dy}{dx} + 1 = 1 - x.$$

Taking square root,

$$\frac{dy}{dx} + 1 = \pm\sqrt{1 - x}, \quad \text{or} \quad \frac{dy}{dx} = -1 \pm (1 - x)^{1/2}$$

which by Formula (V) page 119, integrates to $y = -x \pm \frac{2}{3}(1 - x)^{3/2} + C$. In other words, the differential equation is satisfied by the two algebraic equations

$$y + x - C + \tfrac{2}{3}(1 - x)^{3/2} = 0 \tag{5}$$

$$y + x - C - \tfrac{2}{3}(1 - x)^{3/2} = 0. \tag{6}$$

Since both equations equal zero, their product does likewise, and is also a solution of the differential equation (4), this combined solution being

$$(y + x - C)^2 - \tfrac{4}{9}(1 - x)^3. \tag{7}$$

101. Solution of First Order Equations—Variables Separable.
Case I.   Separation by Simple Rearrangement or Factoring.

When a first order equation contains only the variables x and y, and the derivative $\dfrac{dx}{dy}$, it may often be solved for the derivative to obtain an equation of the form

$$\frac{dy}{dx} = f(x, y)$$

where $f(x, y)$ can be factored into functions $f_1(x)$ of x only and $f_2(y)$ of y only, as

$$\frac{dy}{dx} = f(x, y) = f_1(x)f_2(y)$$

so that we can write

$$\frac{dy}{f_2(y)} = f_1(x)\, dx$$

and integrate separately.

Example. Solve the equation

$$\frac{dy}{dx} = ay. \tag{8}$$

Transposing $\dfrac{dy}{y} = a\,dx.$

Since by Formula (XVI) page 121, the integral of $\dfrac{dy}{y}$ is $\log y$ we have by integration of the above equation $\log y = ax + C.$

Example. Find a formula for the height of water at a given time in a tank of cross-sectional area A when the water is flowing out through an orifice of cross-sectional area a, if the velocity at the orifice, v, is given by the formula $v = k\sqrt{2gh}$, where h is the height of water above the orifice at time t, k is a constant for the system and g is the (constant) acceleration of gravity.

(i) Let dh be the loss of height in time dt.

(ii) Then $A\,dh$ is the loss of volume in time dt.

(iii) Also $av\,dt$ is the flow through the orifice in time dt.

(iv) Substituting $k\sqrt{2gh}$ for v in (iii), we have $ak\sqrt{2gh}\,dt$ for the flow through the orifice in time dt.

Equating (ii) and (iv)

$$-A\,dh = ak\sqrt{2gh}\,dt. \tag{9}$$

Transposing, $\dfrac{dh}{\sqrt{h}} = -\dfrac{a}{A}\,k\sqrt{2g}\,dt$ which gives on integration

$$2\sqrt{h} = -\frac{a}{A}\,k\sqrt{2g}\,t + C$$

or

$$h^{1/2} = C - \frac{k}{2}\frac{a}{A}\sqrt{2g}\,t \tag{10}$$

Here C, the constant of integration, is obviously the square root of the height of liquid when flow started.

Case II: Variables Separable by Substitution.

In many instances, a first order differential equation in which the variables are not readily separable can become so by means of a substitution.

Example. $x\dfrac{dy}{dx} + y + xy^2 = 0$ \hfill (11)

This equation is not directly separable, but by replacing y by $\dfrac{w}{x}$, we have

$$x\,\dfrac{d\left(\dfrac{w}{x}\right)}{dx} + \dfrac{w}{x} + \dfrac{w^2}{x} = 0$$

or

$$x\,\dfrac{\left(\dfrac{xdw - wdx}{x^2}\right)}{dx} + \dfrac{w}{x} + \dfrac{w^2}{x} = 0$$

giving

$$\dfrac{dw}{dx} - \dfrac{w}{x} + \dfrac{w}{x} + \dfrac{w^2}{x} = 0$$

so that

$$\dfrac{dw}{dx} + \dfrac{w^2}{x} = 0.$$

Transposing

$$\dfrac{dx}{x} = -\dfrac{dw}{w^2}.$$

Integrating

$$\log x = C + \log \dfrac{1}{w}$$

Therefore

$$x = Ce^{1/w}$$

Replacing w by xy

$$x = Ce^{1/xy} \tag{12}$$

102. Homogeneous Differential Equations. A differential equation is homogeneous if when written in the form

$$Adx + Bdy = 0, \tag{13}$$

it has for A and B, homogeneous functions of x and y to the same degree. In this case,

$$\dfrac{dy}{dx} = -\dfrac{A}{B} = f\left(\dfrac{y}{x}\right) \tag{13a}$$

and by substituting wx for y, we have

$$\dfrac{dy}{dx} = w + x\,\dfrac{dw}{dx}$$

so that (13a) becomes $w + x\,\dfrac{dw}{dx} = f(w)$

in which the variables are separable, and after integration, replacement of w by $\dfrac{y}{x}$ yields the solution.

A test for this kind of homogeneity is to substitute kx and ky for the terms in x and y respectively. It is homogeneous if this substitution does not change the function, regardless of the value of k.

Example. $$(x^2 + y^2) \frac{dy}{dx} + 2xy = 0. \qquad (14)$$

Substituting kx and ky for x and y gives

$$(k^2x^2 + k^2y^2) \frac{dy}{dx} + 2k^2xy = 0$$

which merely multiplies all terms by k^2, and so does not change the form of the equation. Therefore it is homogeneous, and is solved as follows

$$(x^2 + y^2) \frac{dy}{dx} + 2xy = 0. \qquad (15)$$

Transposing $$\frac{dy}{dx} = -\frac{2xy}{x^2 + y^2}.$$

Substituting wx for y,

$$\frac{d(wx)}{dx} = -\frac{2x^2w}{x^2 + x^2w^2}$$

$$x\frac{dw}{dx} + w = -\frac{2w}{1 + w^2}.$$

Transposing $$x\frac{dw}{dx} = -w - \frac{2w}{1 + w^2}$$

$$x\frac{dw}{dx} = \frac{-w - w^3 - 2w}{1 + w^2}$$

$$x\frac{dw}{dx} = -\frac{3w + w^3}{1 + w^2}$$

$$-\frac{dx}{x} = \frac{1 + w^2}{3w + w^3} dw$$

Integrating $\quad -\log x = \frac{1}{3}\log(3w + w^3) + \log C$

Transposing $\quad \frac{1}{3}\log(3w + w^3) + \log x = -\log C$

Changing from logarithmic form

$$(3w + w^3)^{1/3}x = \frac{1}{C}$$

Cubing both sides $\qquad (3w + w^3)x^3 = \frac{1}{C^3}.$

Substituting $\frac{y}{x}$ for w

$$\left(3\frac{y}{x} + \frac{y^3}{x^3}\right)x^3 = \frac{1}{C^3}$$

$$3x^2y + y^3 = \frac{1}{C^3}$$

Dividing by 3,

$$x^2y + \frac{y^3}{3} = \frac{1}{3C^3} \tag{16}$$

Example. Solve the equation

$$\frac{dy}{dx} = \frac{x+y}{x} \tag{17}$$

Clearing of fractions, $x\,dy = (x+y)dx$

Replacing y by wx, $x(dwx) = (x + wx)dx$

$$x(w\,dx + x\,dw) = (x + wx)dx$$

$$xw\,dx + x^2dw = x\,dx + xw\,dx$$

$$x^2dw = x\,dx$$

$$dw = \frac{dx}{x}$$

Integrating $w = \log x + C$

Replacing w by $\frac{y}{x}$,

$$\frac{y}{x} = \log x + C$$

$$y = x\log x + Cx \tag{18}$$

103. Exact Differential Equations.   An exact differential equation is an equation which is the exact differential of a function that exists; in other words, if there is a function $f(x, y) = 0$, which yields on differentiation the given equation $A\,dx + B\,dy = 0$, then the latter is an exact differential equation. Since the total differential* of $f(x, y)$ is

---

* The total differential (of a function of two variables) is the partial derivative of the second with respect to the first, multiplied by the differential of the first; plus the partial derivative of the first with respect to the second multiplied by the differential of the second. This concept may be generalized for more than two variables. The use of the Greek letter ∂ in expressions such as $\frac{\partial f}{\partial x}$ denotes that they are partial derivatives, in this case with respect to the variable x only. The other variables in the function are treated as constants in such differentiations.

$$\frac{\partial f}{\partial x} dx + \frac{\partial f}{\partial y} dy \qquad (19)$$

then in the exact differential equation,

$$Adx + Bdy = 0,$$

therefore $\qquad A = \dfrac{\partial f}{\partial x} \quad \text{and} \quad B = \dfrac{\partial f}{\partial y}.$

Since $\qquad \dfrac{\partial A}{\partial y} = \dfrac{\partial f}{\partial x \partial y} \quad \text{and} \quad \dfrac{\partial B}{\partial x} = \dfrac{\partial f}{\partial x \partial y},$

therefore $\qquad \dfrac{\partial A}{\partial y} = \dfrac{\partial B}{\partial x}. \qquad (20)$

And conversely, if (20) is true then a function $f(x, y)$ exists for which

$$\frac{\partial f}{\partial x} = A \quad \text{and} \quad \frac{\partial f}{\partial y} = B$$

Therefore equation (20) is a test for the exactness of a differential equation. Since $\dfrac{\partial f}{\partial x} = A$, then f is the integral of A with respect to x, that is, it is

$$f(x, y) = \int^{x} Adx + f_1(y) \qquad (21)$$

because in the integration of a partial derivative the additive term is not a numerical constant, but a function of the other variable or variables. Now differentiate (21) with respect to y, to obtain

$$\frac{\partial f}{\partial y} = \frac{\partial \int^{x} Adx}{\partial y} + \frac{df_1}{dy} = B$$

Then $\qquad \dfrac{df_1}{dy} = B - \dfrac{\partial \displaystyle\int^{x} Adx}{\partial y}.$

which integrates to

$$f_1 = \int \left(B - \frac{\partial \displaystyle\int^{x} Adx}{\partial y} \right) dy. \qquad (22)$$

Substituting (22) in (21) we have

$$f(x, y) = \int^x A dx + \int \left(B - \frac{\partial \int^x A dx}{\partial y} \right) dy = C. \qquad (23)$$

A similar derivation, but with integration of the x-derivative, would yield the equation

$$f(x, y) = \int^y B dy + \int \left(A - \frac{\partial \int^y B dy}{\partial x} \right) dx = C. \qquad (24)$$

Therefore once a differential equation has been found to be exact by the test of equation (20), it may be solved by either (23) or (24).

Example. Solve the differential equation

$$\frac{2xy + 1}{y} dx + \frac{y - x}{y^2} dy = 0.$$

Here $\dfrac{2xy + 1}{y}$ is A and $\dfrac{y - x}{y^2}$ is B.

Hence $\dfrac{\partial A}{\partial y} = \dfrac{\partial \left(\dfrac{2xy + 1}{y} \right)}{\partial y} = -\dfrac{1}{y^2}$

and $\dfrac{\partial B}{\partial x} = \dfrac{\partial \left(\dfrac{y - x}{y^2} \right)}{\partial x} = -\dfrac{1}{y^2}.$

Therefore the equation is exact.

Then $\displaystyle\int^x A dx = \int^x \frac{2xy + 1}{y} dx = \int \left(2x + \frac{1}{y} \right) dx = x^2 + \frac{x}{y}.$

Substituting values for A, B and $\displaystyle\int^x A dx$ in equation (23)

$$f(x, y) = x^2 + \frac{x}{y} + \int \left(\left(\frac{y - x}{y^2} \right) - \frac{\partial \left(x^2 + \frac{x}{y} \right)}{\partial y} \right) dy = C$$

then $f(x, y) = x^2 + \dfrac{x}{y} + \displaystyle\int \left(\left(\frac{1}{y} - \frac{x}{y^2} \right) - \left(-\frac{x}{y^2} \right) \right) dy = C$

[Note that $\dfrac{\partial x^2}{\partial y} = 0$, since terms in x are treated as constant in partial differentiation with respect to y.]

then $$f(x, y) = x^2 + \frac{x}{y} + \int \frac{1}{y} = C$$

giving $x^2 + \dfrac{x}{y} + \log y = C$ as the solution.

Example. Solve the differential equation

$$(x - y)dx - xdy = 0.$$

Test for exactness

$$\frac{\partial(x - y)}{\partial y} = -1$$

$$\frac{\partial(-x)}{\partial x} = -1$$

therefore the equation is exact.

Here $x - y$ is A, $-x$ is B, and $\displaystyle\int^x Adx = \int^x (x - y)dx = \frac{x^2}{2}.$
Then from equation (23)

$$f(x, y) = \frac{x^2}{2} + \int \left(-x - \frac{\partial\left(\dfrac{x^2}{2}\right)}{\partial y} \right) dy = C$$

$$f(x, y) = \frac{x^2}{2} + \int - xdy = C$$

giving $\dfrac{x^2}{2} - xy = C$ as the solution.

104. Equations Made Exact Linear Equations by Integrating Factors. Differential equations occurring in science and industry are not often exact, but every equation of the form

$$f_1(x, y)dx + f_2(x, y)dy = 0$$

can be made exact. A method of procedure is to multiply it by a suitable integrating factor, a function of one or more of the variables which converts the equation to exact form. There is no systematic method of finding integrating factors, but they can sometimes be found by inspection.

Example. Solve the equation

$$ydx - xdy + x^3dx = 0.$$

Here the factor $\dfrac{1}{x^2}$ converts the equation into exact form, for multiplication by it gives

$$\frac{ydx - xdy}{x^2} + xdx = 0$$

and the first term is the derivative of $d\left(\dfrac{x}{y}\right)$, so that we have upon integration

$$\frac{x}{y} + \frac{x^2}{2} = C.$$

This example illustrates the principle of choosing integrating factors to produce expressions having known integrals.

An integrating factor exists for every equation that can be written in the form stated at the beginning of this section; however, such factors are not always readily found.   If the equation can be rearranged in the linear form

$$\frac{dy}{dx} + f(x)y = g(x), \tag{25}$$

the term e^F is always an integrating factor, where F represents the integral of the $f(x)$ term.   On application of e^F, equation (25) becomes

$$\frac{dye^F}{dx} = g(x)e^F$$

which has the solution

$$y = e^{-F}\left(\int e^F g(x)dx + C\right). \tag{26}$$

Example. Solve the equation of Helmholtz for the current in an inductive circuit, which has the form

$$\frac{dI}{dt} + \frac{RI}{L} = \frac{E}{L}, \tag{27}$$

where L is inductance, I, current, t, time, R, resistance, and E, potential. Here $\dfrac{dI}{dt}$ corresponds to $\dfrac{dy}{dx}$, $\dfrac{R}{L}$ corresponds to $f(x)$, and $\dfrac{E}{L}$ corresponds to $g(x)$.   Then

$$F = \int \frac{R}{L} dt = Rt/L.$$

Substituting these values in the general equation (26), we have

$$I = e^{-(Rt/L)}\left(\int \frac{E}{L} e^{(Rt/L)dt} + C\right)$$

$$= \frac{E}{R} + Ce^{-(R/L)t}.$$

105. Equations Reducible to Linear Type. In the preceding section, it was shown how equations of linear type (Eq. 25) have at least one general method of solution. It is important, therefore, that some equations not of linear type may be changed to it by a suitable change of variable.

Example. Solve the equation

$$\frac{dy}{dx} + xy = x. \tag{28}$$

Here $f(x) = x$ and $g(x) = x$, using the symbolism of type equation (25), hence the substitution to be made is $w = ye^F$, where $F = \int x \, dx = \frac{x^2}{2}.$

Therefore
$$w = ye^{x^2/2}$$

from which
$$\frac{dy}{dx} = e^{-x^2/2}\frac{dw}{dx} - xy$$

whence equation (28) becomes

$$\frac{dw}{dx} = xe^{x^2/2}.$$

On integration,
$$w = e^{x^2/2} + C \quad \text{and} \quad y = 1 + Ce^{-x^2/2}. \tag{29}$$

As a more complex example, consider the type equation

$$\frac{dy}{dx} + f_1(x)y = f_2(x)y^n. \tag{30}$$

Division of this equation by $\dfrac{y^n}{1-n}$ gives

$$\frac{1-n}{y^n}\frac{dy}{dx} + (1-n)f_1\frac{1}{y^n-1} = (1-n)f_2.$$

Inspection of this expression shows that the first term is the x-derivative of $\dfrac{1}{y^n-1}$ which also occurs in the second term. Therefore the equation can be made linear by replacing $\dfrac{1}{y^n - 1}$ by w, giving $\dfrac{dw}{dx} +$ $(1 - n)f_1 w = (1 - n)f_2$ which is linear and can then be solved by analogy to equation (25), using the solution (26), to obtain, after replacement of w by y^{1-n},

$$y^{1-n} = e^{-F}\left(C + \int e^F(1 - n)f_2\right)dx \tag{31}$$

where $F = \displaystyle\int (1 - n)f_1 dx$.

106. Differential Equations of Higher Order.

Up to this point, the equations solved have contained only first order derivatives, that is, derivatives such as dy/dx. In other words, they have been first order differential equations. Higher order equations contain higher derivatives such as $\dfrac{d^2x}{dy^2}$, $\dfrac{d^3x}{dy^3}$, etc., the order of the equation being that of the highest derivative.

Let us consider the higher order equation with constant coefficients of the general form

$$a_0 \frac{d^n y}{dx^n} + a_1 \frac{d^{n-1}y}{dx^{n-1}} + \cdots + a_{n-1} \frac{dy}{dx} + a_n y = 0. \tag{32}$$

To solve this equation we need merely recall that the derivative of e^{rx} is

$$\frac{de^{rx}}{dx} = e^{rx}\frac{drx}{dx} = re^{rx}.$$

Moreover

$$\frac{d^2 e^{rx}}{dx^2} = d\frac{\left(\dfrac{de^{rx}}{dx}\right)}{dx} = \frac{dre^{rx}}{dx} = r^2 e^{rx}.$$

Similarly the third, fourth and higher derivatives differ only in that the exponent of the constant factor r increases accordingly.

Therefore, replacing y in Equation (32) by e^{rx} transforms it to

$$f(r) = e^{rx}(a_0 r^n + a_1 r^{n-1} + \cdots + a_{n-1}r + a_n) = 0. \tag{33}$$

Furthermore, if the roots of $f(r)$ are real and different, then the general solution of Equation (32) is

$$y = C_1 e^{r_1 x} + C_2 e^{r_2 x} + \cdots + C_n e^{r_n x}. \tag{34}$$

The term $(a_0 r^n + a_1 r^{n-1} + \cdots a_{n-1}r + a_n)$ in equation (33), which is readily written from (32), is called the auxiliary equation of (32), and enables us to write the solution of (32) directly, as was done in (34).

Example. Solve the equation

$$\frac{d^3y}{dx^3} + 2\frac{d^2y}{dx^2} - \frac{dy}{dx} - 2y = 0.$$

The auxiliary equation is

$$r^3 + 2r^2 - r - 2 = 0.$$

This factors into $(r - 1)(r + 2)(r + 1) = 0.$

Therefore $r = +1, -2,$ or $-1.$

Giving the general solution,

$$y = C_1e^x + C_2e^{-2x} + C_3e^{-x}.$$

Example. Solve the equation

$$\frac{d^2y}{dx^2} + 5\frac{dy}{dx} + 6 = 0.$$

The auxiliary equation is

$$r^2 + 5r + 6 = 0.$$

which factors into

$$(r + 3)(r + 2) = 0.$$

Therefore $r = -3$ or $-2.$

Giving the general solution

$$y = C_1e^{-3x} + C_2e^{-2x}.$$

In stating equation (34) the condition was made that the roots be different. However, they may be the same; as would be the case, for example, if the auxiliary equation were a perfect square. In that case, the independent variable enters as a factor in the solution.

Example. Solve the equation

$$\frac{d^2y}{dx^2} - 6\frac{dy}{dx} + 9y = 0.$$

The auxiliary equation is

$$r^2 - 6r + 9 = 0$$

which factors into

$$(r - 3)^2 = 0.$$

Therefore $r = +3$ (only).

In this case, the general solution is

$$y = C_1 e^{3x} + C_2 x e^{3x}.$$

Up to this point the higher order equations treated have had 0 as their right-hand term.   When it is not zero, but a function of the variables, then the general solution becomes more difficult to find.   In this case it is sometimes feasible to find a particular solution for this equation (that is, a solution specifying certain values of the constants) and also to find a general solution of the corresponding homogeneous equation (called the complimentary function).   Then the sum of the complementary function and the particular solution is a general solution of the equation.

There is no universal method of procedure but one that is often effective is called the *method of undetermined coefficients*, which starts by taking a trial integral with undetermined coefficients, differentiating it, and solving for the coefficients.   The roots of the indicial equation, or the terms on the right-hand side of the original equation (or their derivatives) usually determines the functions in the trial integral.

Example.   Solve the differential equation

$$\frac{d^3y}{dx^3} - 2\frac{d^2y}{dx^2} + \frac{dy}{dx} = e^x + x.$$

First we solve the corresponding homogeneous equation

$$\frac{d^3y}{dx^3} - \frac{d^2x}{dx^2} + \frac{dy}{dx} = 0. \tag{35}$$

Its auxiliary equation is

$$r^3 - 2r^2 + r = 0$$

which factors into

$$(r)(r - 1)(r - 1) = 0.$$

Therefore $r = 0$ or 1 (two roots).
Therefore the general solution is

$$y_1 = C_1 e^{0x} + C_2 e^x + C_3 x e^x$$

or

$$y_1 = C_1 + C_2 e^x + C_3 x e^x. \tag{36}$$

Since $x = 0$ is a root and $x = 1$ is a double root of the auxiliary equation, we use as the trial integral

$$y_2 = Ax^2 e^x + Bx^2 + Cx. \tag{37}$$

Then by repeated differentiation

$$\frac{dy_2}{dx} = Ax^2e^x + 2Axe^x + 2Bx + C$$

$$\frac{d^2y_2}{dx^2} = Ax^2e^x + 4Axe^x + 2Ae^x + 2B$$

$$\frac{d^3y_2}{dx^3} = Ax^2e^x + 6Axe^x + 6Ae^x.$$

Rewriting these terms in column, and multiplying the second order differential by -2 to cancel the terms with the factor of x, we have

$$\frac{d^3y_2}{dx^3} = Ax^2e^x + 2Axe^x \qquad\qquad + 2Bx \qquad + C$$

$$-2\frac{d^2y_2}{dx^2} = -2Ax^2e^x - 8Axe^x - 4Ae^x \qquad\qquad - 4B$$

$$\frac{dy_2}{dx} = Ax^2e^x + 6Axe^x + 6Ae^x$$

$$\frac{d^3y_2}{dx^3} - 2\frac{d^2y_2}{dx^2} + \frac{dy_2}{dx} = 2Ae^x + 2Bx - 4B + C.$$

Then substituting in the original equation (35)

$$2Ae^x + 2Bx - 4B + C = e^x + x.$$

Equating corresponding terms,

$$2Ae^x = e^x, \quad \text{so } A = \tfrac{1}{2}$$

$$2Bx = x, \quad \text{so } B = \tfrac{1}{2}$$

$-4B + C = 0$, and since B is $\tfrac{1}{2}$, so $C = 2$.

Substituting these values in equation (37)

$$y_2 = \tfrac{1}{2}x^2e^x + \tfrac{1}{2}x^2 + 2x \tag{38}$$

This is the particular solution. Since the general solution is the sum of the particular solution, and the general solution (36) of the corresponding homogeneous equation, we have

$$y = \tfrac{1}{2}x^2e^x + \tfrac{1}{2}x^2 + 2x + C_1 + C_2e^x + C_3xe^x.$$

107. Problems for Solution.

1. Solve the differential equation $\dfrac{x\,dy - y\,dx}{x^2 - y^2} = 0$.

2. Solve the differential equation $(y + x)\,dx + x\,dy = 0$.

3. Solve the differential equation $y^2 = x(y - x) \dfrac{dy}{dx}$.

4. Solve the differential equation $2x^2y + y^3 - x^3 \dfrac{dy}{dx} = 0$.

5. If the rate of growth of bacteria in a culture medium is given by the $dN/dt = kN$, where N is the number present at any instant, and k is a constant, find the number present at time t.

6. Solve the differential equation $2xy\, dx - (1 + x^2)\, dy = 0$.

7. Solve the differential equation
$$(x - xy^2)\, dx + (y + x^2y)\, dy = 0.$$

8. Solve the differential equation
$$(y^2 - xy)\, dx + x^2\, dy = 0.$$

9. Solve the differential equation
$$(x^2 + y^2)\, dx - xy\, dy = 0.$$

10. Solve the differential equation
$$(xy - x^2)\, dy - y^2\, dx = 0.$$

11. Solve the differential equation
$$(x - y)\, dx - x\, dy = 0.$$

12. Solve the differential equation
$$\left(5 + \frac{3}{x} - \frac{1}{y}\right) dx + \left(\frac{x}{y^2} - \frac{7}{y} + 2\right) dy = 0.$$

13. Solve the differential equation
$$y\, dx - x\, dy + y^2\, dy = 0.$$

14. Solve the differential equation
$$x\, dx + y\, dy = \sqrt{x^2 + y^2}\, dy.$$

15. Solve the differential equation
$$x \frac{dy}{dx} - (1 - x)y - e^x = 0.$$

16 Solve the differential equation
$$\frac{dy}{dx} - xy - xy^3 = 0$$

17. Solve the differential equation
$$\frac{dy}{dx} + xy = x.$$

18. Solve the differential equation

$$\frac{d^2y}{dx^2} - p^2y = 0, \text{ where } p \text{ is a constant.}$$

19. Solve the differential equation

$$\frac{d^2y}{dx^2} + 12y = 7\frac{dy}{dx}.$$

20. Solve the differential equation

$$\frac{d^4x}{dt^4} - 6\frac{d^3x}{dt^3} + 11\frac{d^2x}{dt^2} - 6\frac{dx}{dt} = e^{-3t}.$$

21. Solve the differential equation

$$\frac{d^2y}{dx^2} - 2\frac{dy}{dx} + y = x - 1.$$

REVIEW PROBLEMS

1. Using the definition of the derivative, $f'(x) = \lim_{h \to 0} \dfrac{f(x+h) - f(x)}{h}$, evaluate $\lim_{h \to 0} \dfrac{f(\frac{1}{2} + h)^6 - f(\frac{1}{2})^6}{h}$.

2. When a gas expands from volume V_0 to volume V_1 it does an amount of work given by $W = \int_{V_0}^{V_1} P\, dV$. If gas expands without transfer of heat energy, the gas law is $P = kV^{-1.4}$ where k is constant. When $\frac{1}{32}$ cubic feet of air under initial pressure of 30 pounds per cubic inch expands to a volume of 1 cubic foot, find the work done.

3. The work done by a force F acting in a direction parallel to the x-axis is given by $\dfrac{dW}{dx} = F$ with the initial conditions $W = 0$ and $x = x_0$. The force acting on a particle P is given by $F = \dfrac{2}{x^3}$. Find the work done in moving the particle from $x = 2$ to $x = 8$.

4. Given the function $f(x) = x^3 - 6x^2 + 9x - 2$:
   a. Find $f'(x)$.
   b. For which values of x is the curve rising?
   c. For which values of x is the curve falling?
   d. Find the coordinates of each relative maximum point and each relative minimum point.
   e. Find $f''(x)$.
   f. For which values of x is the curve concave upward?
   g. For which values of x is the curve concave downward? (A curve is concave upward if $f''(x)$ is positive and concave downward if $f''(x)$ is negative.)
   h. Sketch the curve using the above information.

5. Repeat the entire procedure of problem 4 above if $f(x) = \sin x + \cos x$ where $0 \leqslant x \leqslant 2\pi$.

6. A 12-foot ladder leans on a vertical wall. The lower end is pulled along the horizontal at a constant rate of 3 feet per second. Find the rate at which the top of the ladder moves when the upper end is 8 feet above the floor.

7. The distance s that an object falls in t seconds is given by the equation $s = 400 - 16t^2$. The object falls from a window 400 feet high.
   a. Find its velocity after 1 second.
   b. How long does the object take to reach the ground?
   c. Find its velocity at the instant it hits the ground.

8. A sheet is to contain 72 square inches of printed matter. If the margins at the top and bottom are 2 inches each and the margins at the sides are 1 inch each, find the dimensions of the sheet so that the total area is a minimum.

261

9. An athlete is on an island 4 miles away from a straight beach and her boyfriend lives 3 miles up the beach. She can swim at 4 miles per hour and jog at 6 miles per hour. Find the minimum time required for her to reach her boyfriend.

10. If gas is held at a constant temperature, Boyle's Law states that the pressure P and the volume V are related by the formula $P \cdot V = c$ where c is a constant. Suppose that 250 cubic inches of gas is under a pressure of 100 pounds per square inch. If the pressure is increasing at an instantaneous rate of 10 pounds per square inch per second, find the instantaneous rate of decrease of the volume.

11. The formula for the curvature of the graph of the function $y = f(x)$ is
$$\frac{f''(x)}{[1 + (f'(x))^2]^{3/2}}.$$
Find the curvature of $y = \cos x$ at the point $(0,1)$.

12. If $f(x) = x^{2/3}$,
   a. Find $f'(x)$.
   b. Find a value of x for which $f'(x)$ is not defined.

13. Given the graph of $f(x)$,
   a. graph $f(-x)$.
   b. graph $-f(x)$.

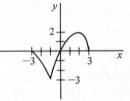

14. If $y = f(x) = \sqrt[3]{x^2 + 1}$, find the new equation which is formed by interchanging x and y in the original equation. This new equation is called the inverse of the given function.
   a. Solve the new equation for y in terms of x.
   b. To which line will the graphs of the original and the inverse function be symmetric?

15. An object is thrown vertically upward such that its distance s from the earth is given by the equation $s = 144t - 16t^2$.
   a. Find the average velocity in the second second of motion.
   b. Find the instantaneous velocity after 3 seconds.
   c. When is the object at rest?
   d. How high will it rise?
   e. With what velocity will the object strike the ground?

16. A function $f(x)$ is defined in the following way:
$$y = f(x) = \begin{cases} x + 3 & \text{for } x \leqslant 2 \\ x^2 + 1 & \text{for } x > 2. \end{cases}$$
   a. Sketch the graph of $f(x)$.
   b. Find $\dfrac{dy}{dx}$.

17. Determine the rate of change of volume with respect to the radius of
    a. a right circular cone of constant height.
    b. a sphere.
18. A farmer has 200 yards of fencing. He wishes to form a rectangular enclosure where a barn is used as a side. Find the dimensions of the enclosure of maximum area.
19. Given $f(x) = x^2 + 2x - 3$,
    a. find $3f(x)$.
    b. find $\dfrac{d(3f(x))}{dx}$.
    c. What effect does multiplying a function by a constant have on its derivative?
    d. find $f(x) + 3$.
    e. find $\dfrac{d(f(x) + 3)}{dx}$.
    f. What effect does adding a constant to a function have on its derivative?
20. Write the equation of the line tangent to the curve $y = 2x^4 - 3x + 7$ at the point $(1,6)$.
21. A manufacturer can make a weekly profit of $25 per item if he makes up to 1,000 items. If he makes more than 1,000 items, then the profit per item decreases by 1¢ per item over 1,000. [For example, if he produces 1,002 items, his profit is $(1,002 × 24.98).] How many items should he produce each week to earn the maximum profit?
22. Bacteria growing in a certain culture increase at a rate proportional to the number present. If there are 10,000 bacteria present initially and 11,000 after 3 hours, how long would it take for the original number of bacteria to double?
23. The side of an equilateral triangle is 18 inches and is increasing at the rate of $1\frac{1}{2}$ inches per hour.
    a. How fast is the perimeter of the triangle increasing?
    b. How fast is the area increasing?
24. A car and a balloon start from the same point at the same instant. The car travels at 48 miles per hour and the balloon rises at the rate of 10 miles per hour. How fast are they separating 30 minutes later?
25. The edge of a regular octahedron is 12 cm. long and is increasing at the rate of 0.2 cm per hour. At what rate is
    a. its surface area increasing?
    b. its volume increasing?
26. An angle is increasing at a constant rate. Show that the tangent of the angle is increasing 8 times as fast as the sine when the angle is $\frac{\pi}{3}$ radians.
27. The acceleration of a moving body is $3t$ where t is in units of time.
    a. If after 6 seconds the velocity is 100 feet per second, express v, the velocity, in terms of t.
    b. If after 3 seconds the distance covered is 90 feet, express s, the distance, in terms of t.
28. A certain firm makes a profit of P each month when it produces x tons of a certain item where $P = 1,500 + 15x^2 - x^3$. Find the most advantageous monthly output for the company.

29. A construction engineer wishes to make an open trough of maximum capacity whose bottom and sides are each 8 inches wide and whose sides have the same slope. What should the distance across the top be?

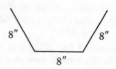

30. In the corner of a field bounded by two perpendicular roads, a spring is located 3 meters from 1 road and 4 meters from the other. How should a straight path be run by this spring and across the corner so as to cut off as little of the field as possible? What is the length of the shortest such road?

31. Find the number which exceeds its square by the greatest possible quantity.

32. Prove that a conical tent of a given capacity will require the least amount of canvas when the height is $\sqrt{2}$ times the base radius.

33. Find the altitude of a right circular cone of maximum volume that is inscribed in a given sphere of radius 6.

34. Given the curve whose equation is $y = x^3 - 3x^2 + 2$, find the coordinates of all points on the curve for which the tangent to the curve at that point is horizontal.

35. For each of the following functions, find $f'(x)$.

   a. $f(x) = \sqrt[3]{2x^2 + 4x}$.

   b. $f(x) = \dfrac{2x + 5}{x^2 - 4}$.

   c. $f(x) = 5x^2(2x^2 + 3)^6$.

36. For each of the following functions, find $f'(x)$.

   a. $f(x) = \sin^2 2x$.

   b. $f(x) = x^2 \cos x$.

   c. $f(x) = \dfrac{e^{2x}}{x}$.

   d. $f(x) = \log_e (7x^3 + 5)$.

37. Evaluate each of the following integrals.

   a. $\int_{\frac{\pi}{6}}^{\frac{\pi}{3}} \sec^3 x \tan x \, dx$.

   b. $\int_0^{13} \sqrt[3]{2x + 1} \, dx$.

   c. $\int_0^1 e^{-x} \, dx$.

38. Find the length of $y = x^{3/2}$ from $x = 0$ to $x = 4$.

39. A circular plate of metal expands by heat so that its radius increases at the rate of 0.03 inches per second. At what rate is the surface area increasing when the radius is 5 inches?

40. Find the coordinates of all points on the curve $y = 4x^3 + 18$ where y is increasing 48 times as fast as x.

41. Integrate:

   a. $\int \sin^2 x \cos^3 x \, dx$.

   b. $\int \tan^2 x \, dx$.

   c. $\int \tan 3x \, dx$.

42. Integrate:

   a. $\int \dfrac{x}{x^2 + 1}\, dx.$

   b. $\int \dfrac{1}{x^2 + 1}\, dx.$

   c. $\int \dfrac{x}{x + 1}\, dx.$

43. Find the area bounded by the x-axis and the lines whose equations are $x - y = 1$ and $3x + y = 15$.

44. Find the area between the curve $xy = 12$ and the line $3x + 2y = 18$.

45. Find the area bounded by the parabola $x = y^2$ and the line $y = x - 2$.

46. Find the area enclosed between the parabolas $y^2 = -4(x - 1)$ and $y^2 = -2(x - 2)$.

47. Find the area bounded by the x-axis and one arch of the curve $y = \sin x$.

48. Find the area between the curve $y = 9 - x^2$ and the x-axis.

49. Find the area bounded by the x-axis, the curve $y = \frac{1}{x}$, the vertical line $x = 1$, and

   a. the vertical line $x = 2$.

   b. the vertical line $x = 5$.

   c. the vertical line $x = k$.

50. Solve the differential equations:

   a. $\dfrac{d^2 y}{dx^2} - \dfrac{dy}{dx} - 6y = 0.$

   b. $\dfrac{d^2 y}{dx^2} - \dfrac{dy}{dx} - 6y = 12.$

ANSWERS TO PROBLEMS

1. $\frac{3}{16}$.

2. $\frac{225}{128}$.

3. $\frac{15}{64}$.

4. a. $3(x^2 - 4x + 3)$.
   b. $x > 3$ or $x < 1$.
   c. $1 < x < 3$.
   d. relative max. $(1, 2)$;
      relative min. $(3, -2)$.
   e. $6x - 12$.
   f. $x > 2$.
   g. $x < 2$.
   h.

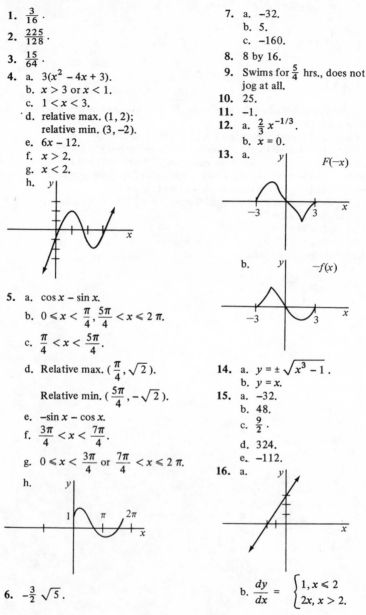

5. a. $\cos x - \sin x$.
   b. $0 \leqslant x < \frac{\pi}{4}, \frac{5\pi}{4} < x \leqslant 2\pi$.
   c. $\frac{\pi}{4} < x < \frac{5\pi}{4}$.
   d. Relative max. $(\frac{\pi}{4}, \sqrt{2})$.
      Relative min. $(\frac{5\pi}{4}, -\sqrt{2})$.
   e. $-\sin x - \cos x$.
   f. $\frac{3\pi}{4} < x < \frac{7\pi}{4}$.
   g. $0 \leqslant x < \frac{3\pi}{4}$ or $\frac{7\pi}{4} < x \leqslant 2\pi$.
   h.

6. $-\frac{3}{2}\sqrt{5}$.

7. a. -32.
   b. 5.
   c. -160.

8. 8 by 16.

9. Swims for $\frac{5}{4}$ hrs., does not jog at all.

10. 25.

11. -1.

12. a. $\frac{2}{3}x^{-1/3}$.
    b. $x = 0$.

13. a.

    b.

14. a. $y = \pm\sqrt{x^3 - 1}$.
    b. $y = x$.

15. a. -32.
    b. 48.
    c. $\frac{9}{2}$.
    d. 324.
    e. -112.

16. a.

    b. $\dfrac{dy}{dx} = \begin{cases} 1, x \leqslant 2 \\ 2x, x > 2. \end{cases}$

266

17. a. $\dfrac{dv}{dr} = 2\pi rh$.

    b. $\dfrac{dv}{dr} = 4\pi r^2$.

18. 50 by 100.
19. a. $3x^2 + 6x - 9$.
    b. $6x + 6$.
    c. Multiplies by a constant.
    d. $x^2 + 2x$.
    e. $2x + 2$.
    f. none.
20. $y = 5x + 1$.
21. 1750.
22. $\dfrac{3 \log_e 2}{\log_e 11}$.

23. a. $\dfrac{9}{2}$.

    b. $\dfrac{27}{2} \sqrt{3}$.

24. $\dfrac{1202}{\sqrt{601}}$.

25. a. $\dfrac{48}{5} \sqrt{3}$.

    b. $\dfrac{144}{5} \sqrt{2}$.

26. $\dfrac{d\theta}{dt} = K$;

    $\dfrac{d(\tan\theta)}{dt} = 4K$;

    $\dfrac{d(\sin\theta)}{dt} = \dfrac{1}{2}K$.

27. a. $V = \dfrac{3}{2}t^2 + 46$.

    b. $S = \dfrac{1}{2}t^3 + 46t - \dfrac{123}{2}$.

28. 10 tons per month for $2,000.
29. 16.
30. a. 6 meters from corner in one direction and 8 meters in the perpendicular direction.
    b. 10.
31. $\dfrac{1}{2}$.

32. $r = \sqrt[3]{\dfrac{3K}{\pi\sqrt{2}}}$;

    $h = \sqrt[3]{\dfrac{6K}{\pi}}$.

33. 8.
34. $(0,2)$, $(2,-2)$.
35. a. $\dfrac{4(x+1)}{3(2x^2+4x)^{2/3}}$.

    b. $\dfrac{-2(x^2+5x+4)}{(x^2-4)^2}$.

    c. $10x(2x^2+3)^5(14x^2+3)$.

36. a. $4 \sin 2x \cos 2x$ or $2 \sin 4x$.
    b. $-x^2 \sin x + 2x \cos x$.

    c. $\dfrac{e^{2x}(2x-1)}{x^2}$.

    d. $\dfrac{21x^2}{7x^3+5}$.

37. a. $\dfrac{1}{27}(72 - 8\sqrt{3})$.

    b. 30.

    c. $1 - \dfrac{1}{e}$.

38. $\dfrac{4}{9}(10\sqrt{10} - 1)$.
39. $.3\,\pi$.
40. $(2,50)$, $(-2,-14)$.
41. a. $\dfrac{1}{3}\sin^3 x - \dfrac{1}{5}\sin^5 x + C$.

    b. $\tan x - x + C$.

    c. $-\dfrac{1}{3}\log_e \cos 3x + C$.

42. a. $\dfrac{1}{2}\log_e(x^2+1) + C$.

    b. $\arctan x + C$.
    c. $x - \log_e(x+1) + C$.
43. 8.
44. $9 - 12 \log_e 2$.
45. $\dfrac{9}{2}$.

46. $\dfrac{8}{3}$.

47. 2.
48. 36.
49. a. $\log_e 2$.
    b. $\log_e 5$.
    c. $\log_e K$.
50. a. $y = C_1 e^{3x} + C_2 e^{-2x}$.
    b. $y = C_1 e^{3x} + C_2 e^{-2x} - 2$.

ANSWERS TO EXERCISES

Following Article 19, Page 24

1. $\left(3x^2 - 4x + \dfrac{1}{2\sqrt{x}}\right) dx.$

2. $\frac{1}{2}(3x^{1/2} - x^{-1/2}) \, dx.$

3. $(x^4 - x^3 + x^2 + x + 1) \, dx.$

4. $-12(7 - 3x)^3 \, dx.$

5. $-18(4 - 2x^3)^2 x^2 \, dx.$

6. $-\dfrac{8 \, dx}{(x - 1)^3}.$

7. $\dfrac{-x^2}{\sqrt[3]{(1 - x^3)^2}} \, dx.$

8. $-\dfrac{x(2x^3 + 15x + 12) \, dx}{(2x^2 + 5)^2}.$

9. $-\dfrac{t \, dt}{(1 + t)^{3/2}}.$

10. $2y \, dy - \dfrac{(4 + 3y) \, dy}{\sqrt{3 - 2y}}.$

11. $4(2x + 7) \, dx.$

12. $\dfrac{9 \, dx}{(x^2 - 6x)^{3/2}}.$

13. $\frac{5}{2}x^{-1/2}.$

14. $-4x^{-2} + 14x^{-3}.$

15. $2x^{-2/3} - 6x^{-1/3}.$

16. $\dfrac{1}{\sqrt{x}} + \dfrac{1}{4x\sqrt{x}}.$

17. $\dfrac{2}{\sqrt{3 + 4x}}.$

18. $-\dfrac{1}{\sqrt[3]{(4 - 3x)^2}}.$

19. $\dfrac{x}{(a^2 - x^2)^{3/2}}.$

20. $-\dfrac{2a}{(a + x)^2}.$

Article 22, Page 35

1. 5.03 cu. in./sec.
2. 10 sq. in./min.
3. (i) Appr. 2 ft./min.
   (ii) Sep. 6 ft./min.
   (iii) After $1\frac{1}{2}$ min.
4. $7\frac{1}{2}$ mi./hr.
5. $4\frac{1}{2}$ mi./hr.
6. $2\frac{4}{9}$ ft./sec.
7. 5 ft./sec.
8. $\frac{5}{2}\sqrt{6} = 6.1$ ft.
9. 283 ft.
10. Sep. $2\frac{1}{2}$ mi./hr.

11. 36 times.
12. $\frac{72}{250}$ cu. in./degree.
13. 4.07 ft./min.
14. 6 ft./sec.
15. $\dfrac{1}{9\pi} = 0.035$ in./sec.
16. 10.6 ft./sec.
17. $25\pi.$
18. $1\frac{3}{4}$ sq. in./min.
19. 10 sq. ft./min.
20. 4.4 mi./min.

Following Article 28, Page 50

1. $2(\cos 2x + \cos x)\, dx.$

2. $(3 \sin x - 2) \sin x \cos x\, dx.$

3. $-3(\cos^2 x \sin x + x^2 \sin x^3)\, dx.$

4. $-\dfrac{\sin \sqrt{1 - t}\, dt}{2\sqrt{1 - t}}.$

5. $(\cos^2 x - \sin^2 x)\, dx.$

6. $(x \cos x + \sin x)\, dx.$

7. $(\tfrac{1}{2} \cos 2\theta - \theta \sin 2\theta)\, d\theta.$

8. $0.$

9. $-\dfrac{x \sin x + \cos x}{x^2}.$

10. $2 \sin x \cos 2x + \cos x \sin 2x.$

11. $\tan 2x \sqrt{\sec 2x}.$

12. $\tan^4 x.$

13. $-\dfrac{2 \sin x}{(1 - \cos x)^2}.$

14. $\cos 2x.$

15. 0.88 rad. or 50.5 deg./min.

16. 0.577 mi.; $\tfrac{1}{3}$ mi./min.

17. $\tfrac{1}{3}$ rad./min.; $.314$ rad./min.

18. $7\tfrac{1}{3}$; $\tfrac{33}{65} = 0.508$ rad./min.

19. Increasing, $1 + \dfrac{1}{\sqrt{2}}$;

decreasing, $\sqrt{3}.$

20. $\dfrac{5k}{32}.$

Following Article 32, Page 58

1. $\dfrac{2}{x^3}.$

2. $-4 \cos 2x.$

3. $12(x^2 - 6x + 8).$

4. $2(x + \sin x).$

5. $\dfrac{d^2 y}{dx^2} = -\cos x = -y.$

6. $\dfrac{2x(3 - 3x + x^2)}{(1 - x)^3}.$

7. $\dfrac{4}{(4 + x^2)^{3/2}}.$

8. $2 \cos x - x \sin x.$

9. $2 \sec^2 x \tan x.$

10. $\dfrac{(2 - x^2) \cos x + 2x \sin x}{x^3}.$

11. $v = 12$ ft./sec.; $a = -18$ ft./sec^2.

12. 72 ft./sec.; 6 ft./sec^2.

13. $\dfrac{1}{2} + \dfrac{\pi}{4} = 1.285$ ft./sec.; 2/sec^2.

14. $11\tfrac{3}{4}$ ft./sec.; $12\tfrac{1}{4}$ ft./sec^2.

15. -0.738 ft./min.; -3.5 ft./sec^2.

16. *a.* $96, -64.$

b. $-32.$

17. $4.$

Following Article 36, Page 75

1. $m = -2; \phi = 116° 34'.$

2. $m = -\tfrac{3}{4}; \phi = 143° 8'.$

3. $m = 0; \phi = 0.$

4. $m = \tfrac{8}{5}; \phi = 57° 2'.$

5. $m = -\tfrac{4}{9}; \phi = 23° 58'.$

6. $x = a.$

7. $m = 0,$ and $1.$

8. $x = \pm 3.$

9. $x = \pm 3.$

10. At $x = 4.$

11. $y = 4x.$

12. $3, 7.$

Following Article 40, Page 86

1. $x = -2$ gives y max. $= 69$;
$x = 3$ gives y min. $= -56.$

2. $x = 6$ gives y min. $= -26$;
$x = -2$ gives y max. $= 28.$

3. $x = 6$ gives y min. $= -348.$

4. $x = 1$ gives y max. $= 3$;
$x = -1$ gives y min. $= -3.$

5. $x = -6$ gives y max. $= -9.$

6. $x = 0$ gives y min. $= 0$;
$x = -2$ gives y max. $= -4.$

7. $x = \tan^{-1}\left(-\frac{1}{2}\right) = 26° 34'$ gives y min. $= -\dfrac{1}{\sqrt{5}} = -0.447$.

8. $x = 180°$ gives y min. $= 0$.

9. $x = 0$ gives no max. or min.; $x = 120°$ gives y min. $= \sqrt{\frac{3}{2}}$.

10. $x = \frac{1}{2}[\tan^{-1}(-8)]$ gives y min. $= \dfrac{1 - \sqrt{65}}{8}$.

Article 43, Page 97

1. $42\frac{2}{3}$; $5\frac{1}{3}$.

2. 8; 12.

3. Depth, 5 in.; base 10 in. square.

4. $H = \frac{3}{4}D$.

5. $R\sqrt{2}$, a square.

6. 13 mi.

8. $2\pi\left(1 - \dfrac{\sqrt{6}}{3}\right)$ rad. $= 66° 14'$.

9. $\dfrac{b}{\sqrt{2}}$ ft.

10. Breadth, $\dfrac{2}{\sqrt{3}}$; depth, $2\sqrt{\dfrac{2}{3}}$ ft.

11. 12.6 mi./hr.

12. 150.

13. 75 miles from town.

14. $\frac{3}{4}$ width of page.

15. 18×24 ft.

16. 40×8 rods.

17. $6\frac{1}{4}$ in.

18. $\dfrac{20}{\sqrt[3]{13}} \times \dfrac{20}{\sqrt[3]{13}} \times \dfrac{5}{2}\sqrt[3]{169}$

   or $8.60 \times 8.60 \times 13.83$.

19. 600 sq. ft.

20. Rectangle, 6×8 ft.

22. About $55°$.

23. $\theta = \dfrac{\pi}{4} + \dfrac{A}{2}$.

24. $\tan\theta = \dfrac{f + \tan\phi - \sec\phi\sqrt{1 + f\cot\phi}}{f\tan\phi - 1}$.

Following Article 48, Page 108

1. $dy = -\dfrac{dx}{x(\log_e x)^2}$.

2. $dy = x^{n-1}(1 + n\log_e x)\,dx$.

3. $dy = \dfrac{2ab\,dx}{a^2x^2 - b^2}$.

4. $dy = b^x e^x(1 + \log_e b)\,dx$.

5. $dy = 9e^{3x}(e^{3x} + 1)^2\,dx$.

6. $dy = x^4 5^x(5 + x\log_e 5)\,dx$.

7. $dy = \dfrac{dx}{x\log_e x}$.

8. $dy = \dfrac{dx}{\sqrt{x^2 - 1}}$.

9. $x = 0$ gives y min. $= 1$.

10. $x = 0$ gives y min. $= 0$.

11. $x = \dfrac{1}{e}$ gives y min. $= -\dfrac{1}{e}$.

12. $x = e$ gives y min. $= e$.

13. y min. $= 2ab$.

14. $x = 0$ gives y max. $= 1$.

Following Article 50, Page 111

1. $dy = 3(x^2 - 1)\,dx$.

2. $dy = \left(\dfrac{1}{a} - \dfrac{a}{x^2}\right)dx$.

3. $dy = \dfrac{a\,dx}{2\sqrt{ax + b}}$.

4. $dy = \dfrac{(a^2 - 2x^2)\,dx}{\sqrt{a^2 - x^2}}$.

5. $ds = abe^{bt}\,dt$.

6. $du = \dfrac{dv}{v}$.

7. $dr = a \cos (a\theta) \, d\theta.$

8. $dy = \cot x \, dx.$

9. $dr = (\cos \theta - \theta \sin \theta) \, d\theta.$

10. $ds = e^t(\cos t - \sin t) \, dt.$

11. $dy = \left(\sqrt{\dfrac{x}{a}} + \sqrt{\dfrac{a}{x}} \right) \dfrac{dx}{2x}.$

12. $du = \dfrac{e^v \, dv}{2\sqrt{e^v + 1}}.$

13. $dy = \dfrac{a^2 \, dx}{(a^2 - x^2)^{3/2}}.$

14. $dy = \dfrac{-a \, dx}{(a + x)\sqrt{a^2 - x^2}}.$

15. $dr = \cos (\tfrac{1}{2}\theta) \, d\theta.$

16. $ds = e^{-at}(b \cos bt - a \sin bt) \, dt.$

17. $dr = -\dfrac{\csc^2 \theta \, d\theta}{2\sqrt{\cot \theta}}.$

18. $dy = \dfrac{3 \, dx}{(6x - 5)(4 - 3x)}.$

Following Article 57, Page 133

1. $\dfrac{5x^4}{4} + \dfrac{2x^{3/2}}{3} - x^2 + C.$

2. $\dfrac{3x^4}{2} - \dfrac{7}{2x^2} + 2x^2 + C.$

3. $\dfrac{e^x}{2} - \dfrac{2x^{5/4}}{5} + C.$

4. $2 \log_e x - \dfrac{10x^{3/2}}{3} + C.$

10. $C - \tfrac{1}{3} \cos 3x + \tfrac{1}{5} \sin 5x + 2 \cos (\tfrac{1}{2}x).$

11. $C + \log_e \left(\dfrac{\sec \theta + \tan \theta}{\cos \theta} \right).$

12. $C + \tfrac{1}{2} \sin^2 \theta.$

13. $C - \tfrac{1}{2}\sqrt{5 - 4x}.$

14. $\dfrac{8x}{9} \sqrt{3x} + C.$

15. $C - \tfrac{1}{6}(3 - 2x)^3.$

16. $C - \dfrac{3}{2(x^2 + 1)}.$

17. $\dfrac{\sin^2 x}{3} + C.$

5. $2ax(\log_e x - 1) + C.$

6. $\dfrac{2^x \log_2 e}{3} + C.$

7. $\dfrac{y}{10} \log_{10} \left(\dfrac{y}{e} \right) + C.$

8. $\tfrac{1}{2}x^2 - x - \log_e x + \dfrac{1}{x} + C.$

9. $\log_e (x^2 + 1) + C.$

18. $\dfrac{\tan^2 ax}{2a} + C.$

19. $C - \dfrac{1}{1 + \tan x}.$

20. $\log_e (x^2 + 5x) + C.$

21. $2 \log_e (4 + \sin x) + C.$

22. $\dfrac{a}{2} (e^{2x/a} - e^{2x/a}) - 2x + C.$

23. $\tfrac{1}{2}e^{(x^2)} + C.$

24. $\dfrac{a^x e^x}{1 + \log_e a} + C.$

25. $C - e^{1/x}.$

Following Article 62, Page 142

1. $\dfrac{9x^{13}}{13} - \dfrac{25x^3}{3} + C.$

2. $\dfrac{8x^7}{7} + 2x^6 + \dfrac{6x^5}{5} + \dfrac{x^4}{4} + C.$

3. $\tfrac{1}{2}x^2 + x + C.$

4. $\tfrac{1}{2}x^2 - x + a \log_e (x - 1) + C.$

5. $\dfrac{x^4}{4} - \dfrac{2x^3}{3} - x - \dfrac{3}{x} + C.$

6. $\tfrac{2}{5}x^{5/2} + \tfrac{4}{7}x^{7/2} - \tfrac{2}{9}x^{9/2} + C.$

7. $\dfrac{\sin ax}{a} + C.$

8. $\sqrt{2x + 5} + C.$

9. $\tfrac{2}{5}(x - 2)\sqrt{(x + 3)^3} + C.$

10. $2e^{\frac{x}{2}} + C.$

11. $\log_e (x^2 + 6x + a) + C.$

12. $e^x(x - 1) + C.$

13. $C - e^{-x}(x^2 + 2x + 2)$.

14. $C + \tan^{-1}\left(\dfrac{x}{3}\right)$.

15. $\log_e\left(\dfrac{x - 1}{x + 1}\right) + C$.

16. $\frac{1}{2}x^2(\log_e x - \frac{1}{2}) + C$.

17. $\frac{1}{4}\log_e(2x^2 - 3) + C$.

18. $2\log_e(x + \sqrt{x^2 - 4}) + C$.

20. $\dfrac{\sqrt{x^2 - 3}}{3x} + C$.

21. $C - \dfrac{1}{x}\sqrt{4 - x^2} - \sin^{-1}(\frac{1}{2}x)$.

22. $\frac{1}{3}x \sin 3x + \frac{1}{9}\cos 3x + C$.

23. $\frac{1}{2}x \tan 2x + \frac{1}{4}\log_e(\cos 2x) + C$.

24. $\dfrac{e^{3x}}{3}\left(x^2 - \dfrac{2x}{3} + \dfrac{2}{9}\right) + C$.

25. $C - \dfrac{x^2 + 2x + 2}{e^x}$.

Following Article 69, Page 161

1. $\frac{5}{2}$.

2. $-\frac{16}{3}$.

3. 2.

4. $200 - \log_{10} e$.

5. $\dfrac{15}{4}a^4 + 3\left(\dfrac{1}{2} - \dfrac{1}{\sqrt[3]{2}}\right)\sqrt[3]{a^2}$.

6. $833\frac{1}{3}$.

7. $-833\frac{1}{3}$.

8. $\log_e 2$.

9. 5.

10. 0.

11. $\dfrac{1}{2}\left(e^2 - \dfrac{1}{e^2}\right)$.

12. $\dfrac{1}{2}\left(e^2 + \dfrac{1}{e^2}\right) - 1$.

13. $\frac{1}{2}\log_e 2$.

14. 1.

15. $-e$.

16. $\frac{16}{3}$.

17. $\frac{98}{3}$.

18. $\dfrac{a^2}{6}$.

19. 1.

20. $\frac{1}{2}(e^\pi + 1)$.

Article 71, Page 173

1. -9.

2. 2.

3. $\frac{2}{3}$.

4. $\frac{8}{3}$.

5. $4\left(\dfrac{1}{e} - 1\right)$.

6. $2(3 - 4e^{-\pi/4})$.

7. \$1333.33.

8. Crosses at $x = 1$; $A = 1$.

11. $2(\pi - \frac{2}{3})a^2$.

12. (a) 1; (b) 2; (c) 3; (d) 4.

Article 73, Page 182

1. $\frac{2}{3}(5\sqrt{5} - 1)$.

2. $\dfrac{2\pi}{3}$.

3. $\frac{19}{27}$.

4. $\log_e(2 + \sqrt{3})$.

5. $\overline{OB} = \sqrt{a^2 + b^2}$.

6. $\displaystyle\int \sqrt{1 + (1 - 4x)^2}\, dx = \frac{1}{8}\left[2\sqrt{2} + \log_e\left(\dfrac{\sqrt{2} + 1}{\sqrt{2} - 1}\right)\right]$.

7. Ellipse; semi-axes $3\sqrt{2}$, $2\sqrt{2}$.

8. $A = 12\pi$.

9. $s = 8a$.

10. $s = \frac{1}{2}a(e^{x/a} - e^{-x/a})$.

Article 89, Page 225

1. (a) 80; (b) 100 ft./sec.
2. (a) 456 ft.; (b) 172 ft./sec.
3. 200 ft.
4. 33.1 deg.
5. $v = Ve^{-kt}$.
6. 4.5 ft.
7. $4\sqrt{10} = 12.65$ ft.
8. (a) 12.2 lb.; (b) 8.15 ft.
9. $k = \dfrac{1}{t} \log_e \left(\dfrac{a}{a-x} \right)$.
10. 586 lb.
11. $\dfrac{ds}{dx} = -\dfrac{s}{v}$; $s = s_0 e^{-x/v}$.
12. 6931 gal.
13. 92 sec.
14. $y = v(1 - e^{-x/v})$.

15. 555 ft.
16. 0.958 A_0 remains.
17. $r = r_0 - kt$.
18. 64 lb.
19. 13.5 days.
20. 135,000.
21. 0.0878 of original.
22. $\log_e \left(\dfrac{P}{P_0} \right) = 5343 \left(\dfrac{1}{T_0} - \dfrac{1}{T} \right)$.
23. $R = 60,600$ ft. $= 11.5$ mi.;
    $H = 17,460$ ft. $= 3.31$ mi.;
    $T = 46.6$ sec. $= 0.777$ min.;
    $y = \dfrac{x}{\sqrt{3}} - (9.55 \times 10^{-6})x^2$.
24. $a = 72$; 862 sq. mi.
25. $n = 13.05$ rev./sec. $= 783$
    R.P.M.

Article 107, Page 258

1. $y = Cx$.
2. $xy + \frac{1}{2}x^2 = C$.
3. $y = Ce^{y/x}$.
4. $y = Cx\sqrt{x^2 + y^2}$.
5. $N = Ce^{kt}$
6. $y = C(1 + x^2)$.
7. $1 + x^2 = C^2(1 - y^2)$.
12. $5x + 3 \log x - \dfrac{x}{y} - 7 \log y + 2y = C$.
13. $\dfrac{x}{y} + y = C$.
14. $x^2 = 2Cy + C^2$.
15. $y = e^x(Cx - 1)$.
20. $x = a_0 + a_1 e^t + a_2 e^{2t} + a_3 e^{3t} + \frac{1}{360} e^{-3t}$.
21. $y = C_1 e^x + C_2 x e^x + x + 1$.

8. $\log x - \dfrac{x}{y} = C$.
9. $\log x - \dfrac{y^2}{2x^2} = C$.
10. $\dfrac{y}{x} - \log y = C$.
11. $\dfrac{x^2}{2} - xy = C$.
16. $e^{x^2} = Cy^2 - e^{x^2}y^2$
17. $y = 1 - Ce^{-x^2/2}$.
18. $y = C_1 e^{px} + C_2 e^{-px}$.
19. $y = C_1 e^{3x} + C_2 e^{4x}$.

INTEGRALS

$$a+bx$$

1. $\int \dfrac{dx}{x(a+bx)} = -\dfrac{1}{a}\log\dfrac{a+bx}{x}+C.$

2. $\int \dfrac{dx}{x^2(a+bx)} = -\dfrac{1}{ax}+\dfrac{b}{a^2}\log\dfrac{a+bx}{x}+C.$

3. $\int \dfrac{dx}{x(a+bx)^2} = \dfrac{1}{a(a+bx)}-\dfrac{1}{a^2}\log\dfrac{a+bx}{x}+C.$

4. $\int \dfrac{x\,dx}{(a+bx)^3} = \dfrac{1}{b^2}\left[-\dfrac{1}{a+bx}+\dfrac{a}{2(a+bx)^2}\right]+C.$

$$\sqrt{a+bx}$$

5. $\int x\sqrt{a+bx}\,dx = -\dfrac{2(2a-3bx)\sqrt{(a+bx)^3}}{15b^2}+C.$

6. $\int x^2\sqrt{a+bx}\,dx = \dfrac{2(8a^2-12abx+15b^2x^2)\sqrt{(a+bx)^3}}{105b^3}+C.$

7. $\int \dfrac{x\,dx}{\sqrt{a+bx}} = -\dfrac{2(2a-bx)\sqrt{a+bx}}{3b^2}+C.$

8. $\int \dfrac{x^2\,dx}{\sqrt{a+bx}} = \dfrac{2(8a^2-4abx+3b^2x^2)\sqrt{a+bx}}{15b^3}+C.$

$$\sqrt{x^2+a^2}$$

9. $\int \sqrt{x^2+a^2}\,dx = \dfrac{x}{2}\sqrt{x^2+a^2}+\dfrac{a^2}{2}\log\,(x+\sqrt{x^2+a^2})+C.$

10. $\int \sqrt{(x^2+a^2)^3}\,dx = \dfrac{x}{8}(2x^2+5a^2)\sqrt{x^2+a^2}+\dfrac{3a^4}{8}\log\,(x+\sqrt{x^2+a^2})+C.$

11. $\int x^2\sqrt{x^2+a^2}\,dx = \dfrac{x}{8}(2x^2+a^2)\sqrt{x^2+a^2}-\dfrac{a^4}{8}\log\,(x+\sqrt{x^2+a^2})+C.$

12. $\int \dfrac{dx}{\sqrt{x^2+a^2}} = \log\,(x+\sqrt{x^2+a^2})+C.$

13. $\int \dfrac{dx}{\sqrt{(x^2+a^2)^3}} = \dfrac{x}{a^2\sqrt{x^2+a^2}}+C.$

14. $\int \dfrac{x^2\,dx}{\sqrt{x^2+a^2}} = \dfrac{x}{2}\sqrt{x^2+a^2}-\dfrac{a^2}{2}\log\,(x+\sqrt{x^2+a^2})+C.$

15. $\int \dfrac{x^2\,dx}{\sqrt{(x^2+a^2)^3}} = -\dfrac{x}{\sqrt{x^2+a^2}}+\log\,(x+\sqrt{x^2+a^2})+C.$

16. $\int \dfrac{dx}{x\sqrt{x^2+a^2}} = \dfrac{1}{a}\log\dfrac{x}{a+\sqrt{x^2+a^2}}+C.$

17. $\int \dfrac{dx}{x^2\sqrt{x^2+a^2}} = -\dfrac{\sqrt{x^2+a^2}}{a^2x}+C.$

18. $\int \dfrac{\sqrt{x^2+a^2}}{x}\,dx = \sqrt{a^2+x^2}-a\log\dfrac{a+\sqrt{a^2+x^2}}{x}+C.$

19. $\int \dfrac{\sqrt{x^2+a^2}}{x^2}\,dx = -\dfrac{\sqrt{x^2+a^2}}{x}+\log\,(x+\sqrt{x^2+a^2})+C.$

$$\sqrt{x^2-a^2}$$

20. $\int \sqrt{x^2-a^2}\,dx = \dfrac{x}{2}\sqrt{x^2-a^2}-\dfrac{a^2}{2}\log\,(x+\sqrt{x^2-a^2})+C.$

21. $\int \sqrt{(x^2-a^2)^3}\,dx = \dfrac{x}{8}(2x^2-5a^2)\sqrt{x^2-a^2}+\dfrac{3a^4}{8}\log\,(x+\sqrt{x^2-a^2})+C.$

22. $\int x^2\sqrt{x^2-a^2}\,dx = \dfrac{x}{8}(2x^2-a^2)\sqrt{x^2-a^2}-\dfrac{a^4}{8}\log\,(x+\sqrt{x^2-a^2})+C.$

23. $\int \dfrac{dx}{\sqrt{x^2-a^2}} = \log\,(x+\sqrt{x^2-a^2})+C.$

24. $\int \dfrac{dx}{\sqrt{(x^2-a^2)^3}} = -\dfrac{x}{a^2\sqrt{x^2-a^2}}+C.$

25. $\int \dfrac{x^2\,dx}{\sqrt{x^2-a^2}} = \dfrac{x}{2}\sqrt{x^2-a^2}+\dfrac{a^2}{2}\log\,(x+\sqrt{x^2-a^2})+C.$

26. $\int \dfrac{dx}{x^2\sqrt{x^2-a^2}} = \dfrac{\sqrt{x^2-a^2}}{a^2x}+C.$

27. $\int \dfrac{\sqrt{x^2-a^2}}{x}\,dx = \sqrt{x^2-a^2} - a\cos^{-1}\dfrac{a}{x} + C.$

28. $\int \dfrac{\sqrt{x^2-a^2}}{x^2}\,dx = -\dfrac{\sqrt{x^2-a^2}}{x} + \log\,(x+\sqrt{x^2-a^2}) + C.$

$$\sqrt{a^2-x^2}$$

29. $\int \sqrt{a^2-x^2}\,dx = \dfrac{x}{2}\sqrt{a^2-x^2} + \dfrac{a^2}{2}\sin^{-1}\dfrac{x}{a} + C.$

30. $\int \sqrt{(a^2-x^2)^3}\,dx = \dfrac{x}{8}(5a^2-2x^2)\sqrt{a^2-x^2} + \dfrac{3a^4}{8}\sin^{-1}\dfrac{x}{a} + C.$

31. $\int x^2\sqrt{a^2-x^2}\,dx = \dfrac{x}{8}(2x^2-a^2)\sqrt{a^2-x^2} + \dfrac{a^4}{8}\sin^{-1}\dfrac{x}{a} + C.$

32. $\int \dfrac{dx}{\sqrt{(a^2-x^2)^3}} = \dfrac{x}{a^2\sqrt{a^2-x^2}} + C.$

33. $\int \dfrac{x^2\,dx}{\sqrt{a^2-x^2}} = -\dfrac{x}{2}\sqrt{a^2-x^2} + \dfrac{a^2}{2}\sin^{-1}\dfrac{x}{a} + C.$

34. $\int \dfrac{dx}{x\sqrt{a^2-x^2}} = \dfrac{1}{a}\log\dfrac{x}{a+\sqrt{a^2-x^2}} + C.$

35. $\int \dfrac{dx}{x^2\sqrt{a^2-x^2}} = -\dfrac{\sqrt{a^2-x^2}}{a^2x} + C.$

36. $\int \dfrac{\sqrt{a^2-x^2}}{x}\,dx = \sqrt{a^2-x^2} - a\log\dfrac{a+\sqrt{a^2-x^2}}{x} + C.$

37. $\int \dfrac{\sqrt{a^2-x^2}}{x^2}\,dx = -\dfrac{\sqrt{a^2-x^2}}{x} - \sin^{-1}\dfrac{x}{a} + C.$

$$\sqrt{2ax-x^2}$$

38. $\int \dfrac{dx}{\sqrt{2ax-x^2}} = \text{vers}^{-1}\dfrac{x}{a} + C.$

39. $\int \sqrt{2ax-x^2}\,dx = \dfrac{1}{2}(x-a)\sqrt{2ax-x^2} + \dfrac{a^2}{2}\text{vers}^{-1}\dfrac{x}{a} + C.$

40. $\int x\sqrt{2ax-x^2}\,dx = -\dfrac{3a^2+ax-2x^2}{6}\sqrt{2ax-x^2} + \dfrac{a^3}{2}\text{vers}^{-1}\dfrac{x}{a} + C.$

41. $\displaystyle\int \frac{dx}{x\sqrt{2ax-x^2}} = -\frac{\sqrt{2ax-x^2}}{ax}+C.$

42. $\displaystyle\int \frac{x\,dx}{\sqrt{2ax-x^2}} = -\sqrt{2ax-x^2}+a\,\mathrm{vers}^{-1}\frac{x}{a}+C.$

43. $\displaystyle\int \frac{\sqrt{2ax-x^2}}{x}\,dx = \sqrt{2ax-x^2}+a\,\mathrm{vers}^{-1}\frac{x}{a}+C.$

44. $\displaystyle\int \frac{x^2\,dx}{\sqrt{2ax-x^2}} = -\frac{x+3a}{2}\sqrt{2ax-x^2}+\frac{3a^2}{2}\,\mathrm{vers}^{-1}\frac{x}{a}+C$

45. $\displaystyle\int \frac{\sqrt{2ax-x^2}}{x^2}\,dx = -\frac{2\sqrt{2ax-x^2}}{x}-\mathrm{vers}^{-1}\frac{x}{a}+C.$

46. $\displaystyle\int \sin^2 x\,dx = \tfrac{1}{2}x-\tfrac{1}{4}\sin 2x+C.$

47. $\displaystyle\int \sin^n x\,dx = -\frac{\sin^{n-1}x\cos x}{n}+\frac{n-1}{n}\int \sin^{n-2}x\,dx+C.$

48. $\displaystyle\int \cos^2 x\,dx = \tfrac{1}{2}x+\tfrac{1}{4}\sin 2x+C.$

49. $\displaystyle\int \cos^n x\,dx = \frac{\cos^{n-1}x\sin x}{n}+\frac{n-1}{n}\int \cos^{n-2}x\,dx+C.$

50. $\displaystyle\int \cos^m x\sin^n x\,dx = \frac{\cos^{m-1}x\sin^{n+1}x}{m+n}+\frac{m-1}{m+n}\int \cos^{m-2}x\sin^n x\,dx+C.$

51. $\displaystyle\int \cos^m x\sin^n x\,dx = -\frac{\sin^{n-1}x\cos^{m-1}x}{m+n}+\frac{n-1}{m+n}\int \cos^m x\sin^{n-2}x\,dx+C.$

52. $\displaystyle\int \tan^n x\,dx = \frac{\tan^{n-1}x}{n-1}-\int \tan^{n-2}x\,dx+C.$

53. $\displaystyle\int \cot^n x\,dx = -\frac{\cot^{n-1}x}{n-1}-\int \cot^{n-2}x\,dx+C.$

54. $\displaystyle\int e^{ax}\sin nx\,dx = \frac{e^{ax}(a\sin nx-n\cos nx)}{a^2+n^2}+C.$

55. $\displaystyle\int e^{ax}\cos nx\,dx = \frac{e^{ax}(n\sin nx+a\cos nx)}{a^2+n^2}+C.$

56. $\displaystyle\int xe^{ax}\,dx = \frac{e^{ax}(ax-1)}{a^2}+C.$

57. $\int x^n e^{ax} \, dx = \dfrac{x^n e^{ax}}{a} - \dfrac{n}{a} \int x^{n-1} e^{ax} \, dx + C.$

58. $\int x \sin x \, dx = \sin x - x \cos x + C.$

59. $\int x \sin^n x \, dx = \dfrac{\sin^{n-1} x(\sin x - nx \cos x)}{n^2} + \dfrac{n-1}{n} \int x \sin^{n-2} x \, dx + C.$

60. $\int x \cos x \, dx = \cos x + x \sin x + C.$

61. $\int x \cos^n x \, dx = \dfrac{\cos^{n-1} x(\cos x + nx \sin x)}{n^2} + \dfrac{n-1}{n} \int x \cos^{n-2} x \, dx + C.$

62. $\int x^n \log x \, dx = x^{n+1} \left[\dfrac{\log x}{n+1} - \dfrac{1}{(n+1)^2} \right] + C.$

63. $\int x^m \log^n x \, dx = \dfrac{x^{m+1}}{m+1} \log^n x - \dfrac{n}{m+1} \int x^m \log^{n-1} x \, dx + C.$

INDEX